Atlas of Spores of Selected Genera and Species of Streptomycetaceae

Atlas of Spores
of Selected Genera and Species
# of Streptomycetaceae

By

W. Kuryłowicz
State Institute of Hygiene, Warsaw

G. F. Gause
Institute for New Antibiotics, Moscow

Alina Paszkiewicz
State Institute of Hygiene, Warsaw

Tatiana P. Preobrazhenskaya
Institute for New Antibiotics, Moscow

Irena Wojewódzka
State Institute of Hygiene, Warsaw

Tatiana S. Maximova
Institute for New Antibiotics, Moscow

Wanda Woźnicka
State Institute of Hygiene, Warsaw

W. Brzosko
State Institute of Hygiene, Warsaw

K. Malinowski
State Institute of Hygiene, Warsaw

S. T. Williams
Hartley Botanical Laboratories
University of Liverpool

1971 University Park Press · Baltimore · London · Tokyo

First published in Poland by Państwowy Zakład Wydawnictw Lekarskich (Polish Medical Publishers), Warsaw

Published in the United States of America and Canada by University Park Press, Baltimore, Maryland

Library of Congress Catalog Card Number LC 85—160168
International Standard Book Number (ISBN) 0—8391—0585—1

Printed in Poland

Contents

Introduction

Actinomycetes belong to the microorganisms producing most of the antibiotics applied in medicine and many other compounds of practical importance. The taxonomy and the classification of Actinomycetes have made many difficulties to investigators of these organisms, because no uniform and commonly applied system of classification has been established.

The first descriptions of Actinomycetes were based on a small number of criteria and were limited to few characteristics. After discovery of their antibiotic properties the taxonomic descriptions comprised a broad range of morphological features and physiological properties. Studies made by Kriss and others in 1945 using electron microscopy revealed a various structure of the surface of spores and allowed for the introduction of a new criterion into taxonomy of these organisms. Studies of Küster (1953) demonstrated that many processes in the form of spines or hair were present on the surface of some spores. Studies with the aid of electron microscope performed by Grein (1955), Flaig and others (1955), Baldacci and Grein (1955), Ettlinger and others (1958), Preobrazhenskaya and others (1959, 1960, 1965), Hopwood (1961), Tresner and others (1960, 1961), Dietz and Mathews (1962), Rancourt and Lechevalier (1963, 1964), Lechevalier and Holbert (1965), Yajima (1965) and Lechevalier and others (1966) showed a varied shape of spores and four types of the surface configuration: smooth, warty, spiny and hairy.

In spite of many observations of the spores in electron microscope, no paper has been published which would summarize all previous observations, thus presenting all the forms and types of spore surface configuration in different genera of Actinomycetes. Okami, in 1965, was the first to make an attempt of such presentation.

The presented material was obtained in three research centres, namely in the State Institute of Hygiene, Warsaw, Poland, the Institute for Search of New Antibiotics, Academy of Medical Sciences, Moscow, USSR, and Hartley Botanical Laboratories, University of Liverpool, England. Photographs of 62 species of Actinomycetes included in the family of Streptomycetaceae are presented. Among others, photographs of the ultrathin sections of spores of some Actinomycetes strains are included.

Investigating application of numerical taxonomy to Streptomyces Woźnicka (1967), Gyllenberg and others (1967) and Kuryłowicz and others (1970) tested 150 strains, belonging to the genus Streptomyces. Some of these were subjected to the ultrastructural studies.

Most attention has been paid to the genus Streptomyces. As for the nomenclature, the strains isolated by Soviet investigators are named „Actinomyces", the term used by Soviet authors. For species isolated in other centres, the original generic name „Streptomyces" is also used.

Actinomycetes belonging to the genus Streptomyces were cultured at the temperature of 28° during 7 to 10 days. Strains included into other genera were cultured at the same temperature for 10 to 20 days except for Actinobifida chromogena which was incubated at 56° for 18 days.

The following media were used: mineral medium No. 1 and No. 2 prepared as described in the manual of Gause and others (1957) and yeast-malt and oat-meal media as reported in the *Methods Manual* by Shirling and Gottlieb (1964).

Spores for electron microscope examination were unstained and mounted of the 150 mesh copper grids covered with formvar film using imprint technique. Formvar films were prepared from 0,5% formvar solution in ethylene dichloride. Some of the examined grids were prepared using shadow-casting technique. The carbon replicas were prepared according to the method described by Preobrazhenskaya and others (1965).

For ultrastructural studies the material was fixed in 4% buffered glutaraldehyde pH 7,2 and postfixed in 1% osmium tetroxide pH 7,2 (Sabatini and others 1963). Material was dehydrated in ethanol and embedded in Epon 812. Ultrathin sections were cut on UT-1 and UT-2 Porter-Blum ultramicrotomes and double stained with uranyl acetate and lead citrate (Echlin, 1964).

Material for scanning electron microscope was prepared according to the method described by Williams and Davies (1967). Scanning electron microscope photographs of spores were taken on Stereoscan Cambridge Instrument Company using Ilford H.P.3 film in Hartley Botanical Laboratories, University of Liverpool, England.

Photographs of spores and ultrathin sections of spores were taken on JEM 6c electron microscope in State Institute of Hygiene in Warsaw, Poland, using Foton and ORWO electron microscope plates.

Carbon replica photographs were taken on EMB-100 A electron microscope in Institute for Search of New Antibiotics, Moscow, USSR.

1

Genus Streptomyces (Actinomyces)

Figures 1—79

Actinomyces (Streptomyces) atrovirens Maximova 1968
INA 300

Spore surface: hairy long
Mineral medium No. 1
Chromium shadow-casting technique

Figure 1
10 000 ×

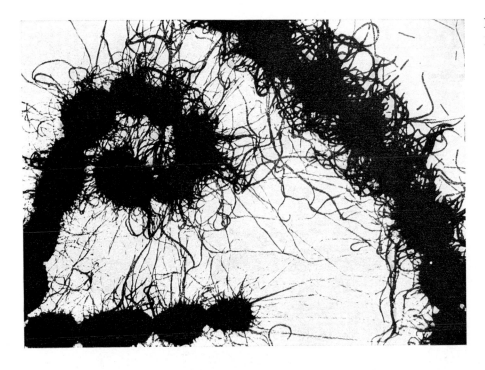

Figure 2
34 000 ×

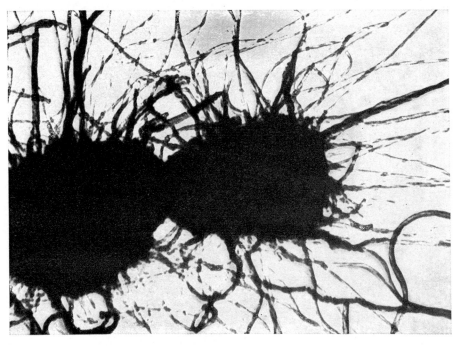

Actinomyces (Streptomyces) atrovirens Maximova 1968 INA 300

Spore surface: hairy long
Mineral medium No. 1

Figure 3
16 000 ×

Figure 4
40 000 ×

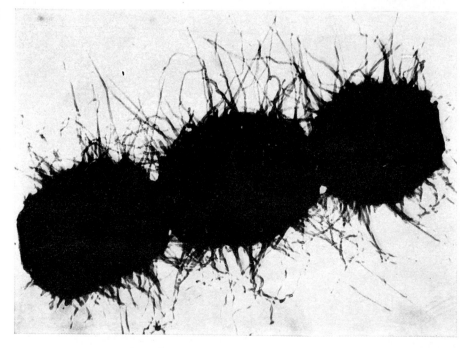

Actinomyces (Streptomyces) atrovirens Maximova 1968 INA 300

Spore surface: hairy long
Mineral medium No. 1
Carbon replica

Figure 5
80 000 $\times$

Actinomyces (Streptomyces) acrimycini Gause et al. 1957 INA 7699

Spore surface: hairy long
Mineral medium No. 1

Figure 6
10 000 ×

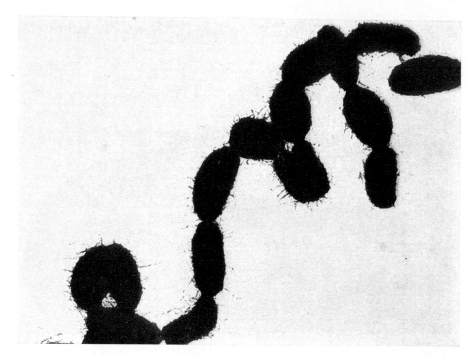

Figure 7
40 000 ×

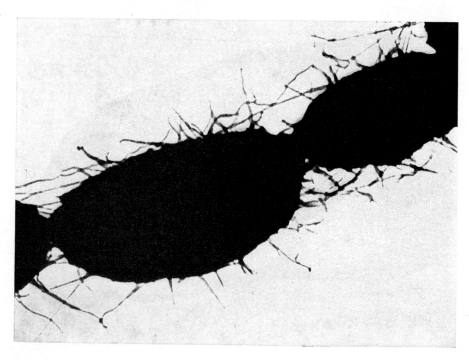

Actinomyces (Streptomyces) flavoviridis Krassilnikov 1941
INA 2314/53

Spore surface: hairy intermediate long
Mineral medium No. 1

Figure 8
16 000 ×

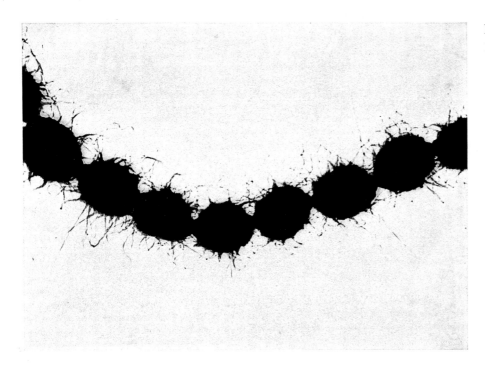

Figure 9
40 000 ×

Streptomyces (Actinomyces) pilosus Ettlinger et al. 1958
ISP 5097

Spore surface: hairy intermediate long
Mineral medium No. 1

Figure 10
16 000 ×

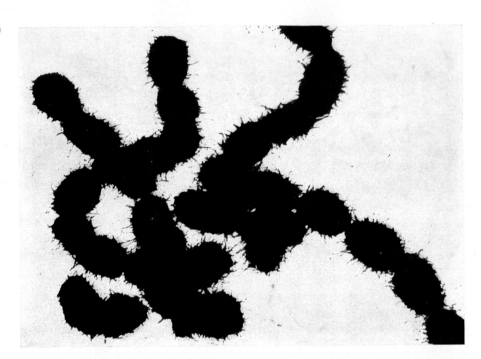

Figure 11
40 000 ×

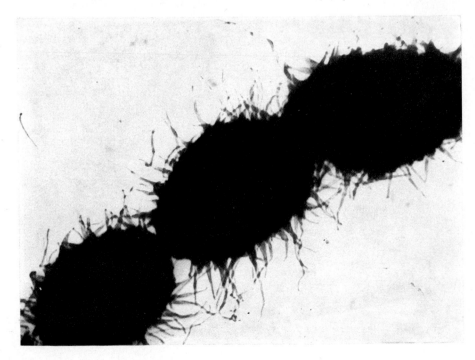

Actinomyces (Streptomyces) glaucescens Gause et Preobrazhenskaya 1957, INA 13886

Spore surface: hairy intermediate long
Mineral medium No. 1

Figure 12
16 000 ×

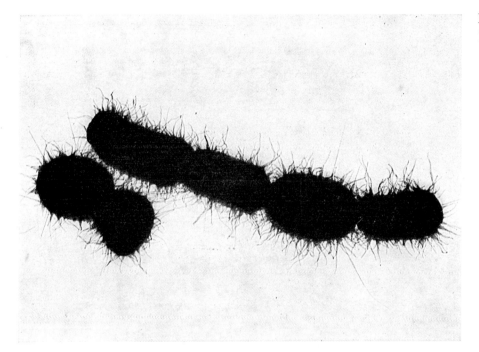

Figure 13
40 000 ×

Streptomyces (Actinomyces) prasinopilosus Ettlinger et al. 1958, ISP 5098

Spore surface: hairy intermediate long
Mineral medium No. 1

Figure 14
10 000 ×

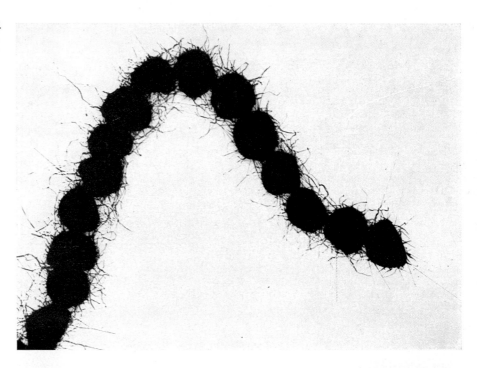

Figure 15
40 000 ×

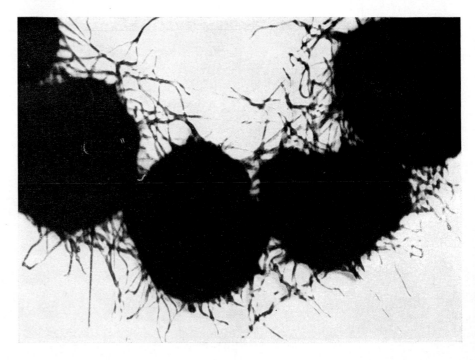

Actinomyces (Streptomyces) nordicus Preobrazhenskaya 1966, INA 1943

Spore surface: hairy intermediate long
Mineral medium No. 1

Figure 16
16 000 ×

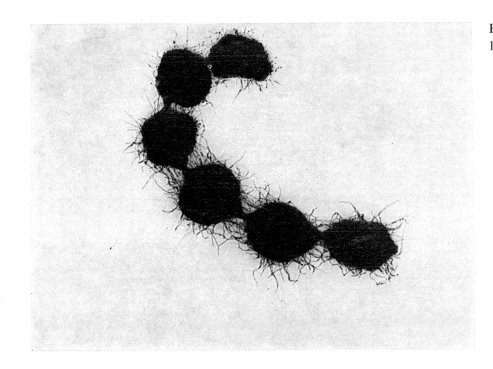

Figure 17
40 000 ×

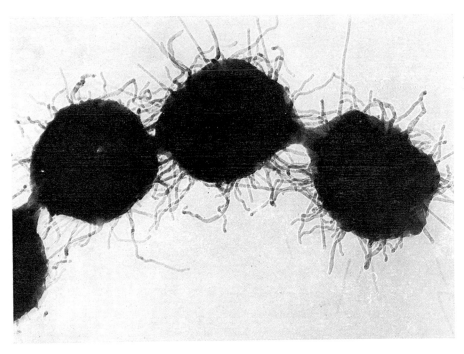

Actinomyces (Streptomyces) griseomycini Gause et al. 1957 INA 7764

Spore surface: hairy short
Mineral medium No. 1

Figure 18
10 000 ×

Figure 19
40 000 ×

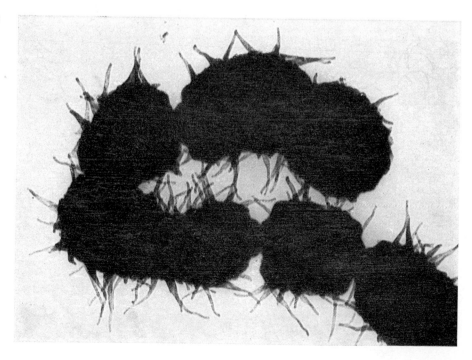

Actinomyces (Streptomyces) viridiviolaceus Gause et al. 1957
INA 313

Spore surface: hairy short
Oatmeal medium

Figure 20
12 000 ×

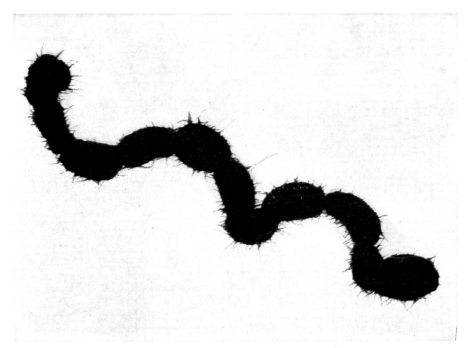

Figure 21
40 000 ×

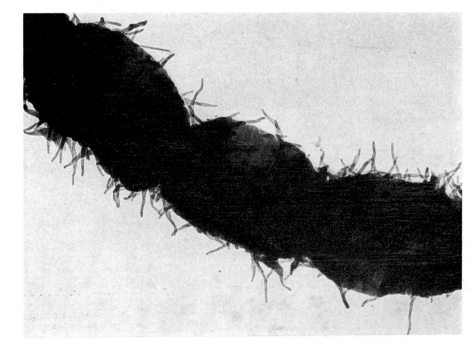

Actinomyces (Streptomyces) longoechinatus Maximova 1968 INA 7332

Spore surface: hairy to spiny
Mineral medium No. 1

Figure 22
16 000 ×

Figure 23
40 000 ×

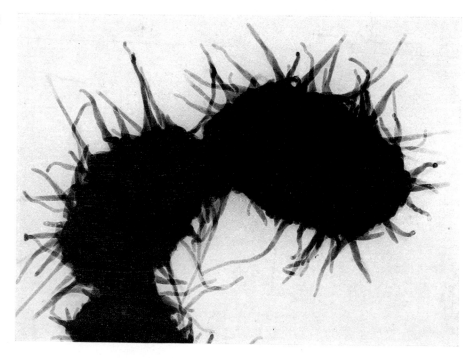

Actinomyces (Streptomyces) flavochromogenes var. heliomycini Gause et al. 1958, INA 2915

Spore surface: hairy to spiny
Mineral medium No. 1

Figure 24
16 000 ×

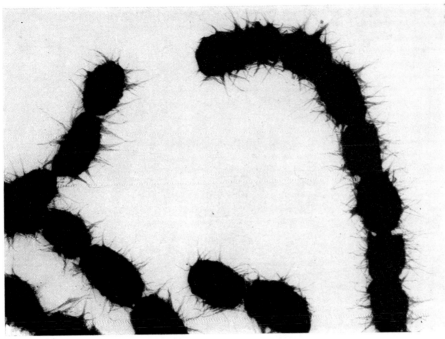

Figure 25
40 000 ×

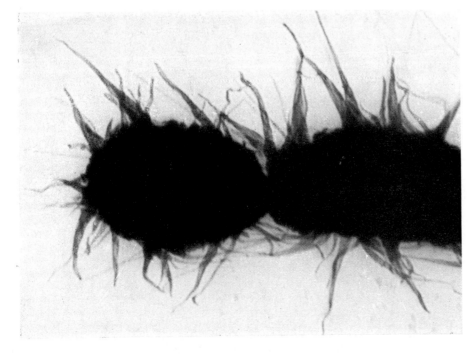

Streptomyces (Actinomyces) viridosporus Parke et al. ISP 5243

Spore surface: hairy to spiny
Mineral medium No. 1

Figure 26
16 000 ×

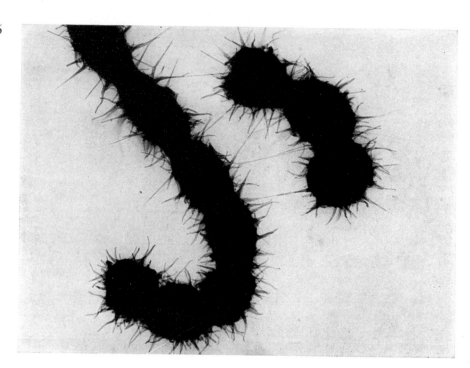

Figure 27
40 000 ×

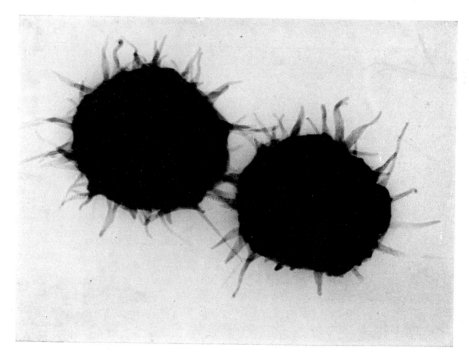

Streptomyces (Actinomyces) fasciculatus McCormick et al. 1953, ISP 5054

Spore surface: spiny long
Mineral medium No. 1

Figure 28
16 000 $\times$

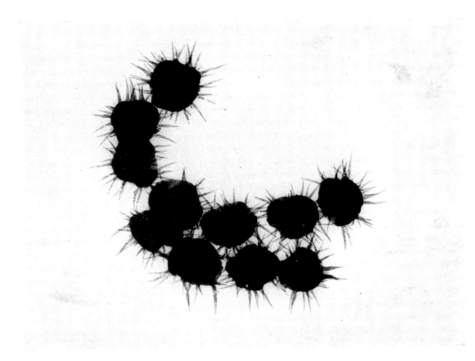

Figure 29
40 000 $\times$

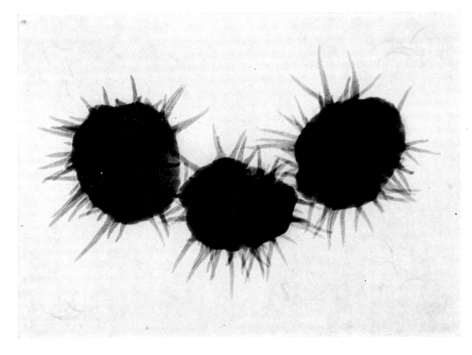

Streptomyces (Actinomyces) thermotolerans Pagano et al. 1959, ISP 5227

Spore surface: spiny intermediate long
Mineral medium No. 1

Figure 30
10 000 ×

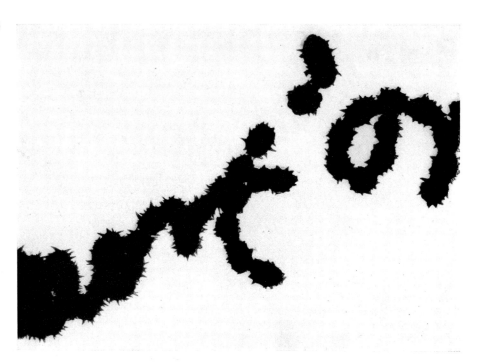

Figure 31
40 000 ×

Actinomyces (Streptomyces) coerulescens Gause et Preobrazhenskaya 1957, INA 1581

Spore surface: spiny intermediate long
Mineral medium No. 1

Figure 32
16 000 ×

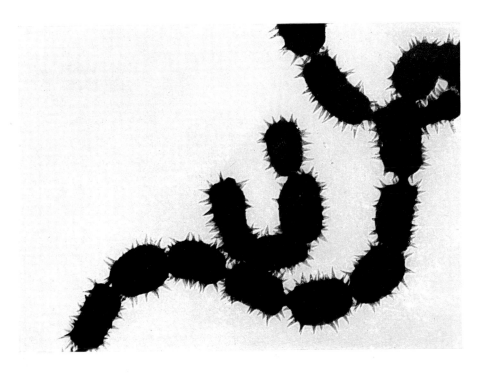

Figure 33
40 000 ×

Actinomyces (Streptomyces) coerulescens Gause et Preobrazhenskaya 1957, INA 1581

Spore surface: spiny intermediate long
Mineral medium No. 1
Carbon replica

Figure 34
25 000 ×

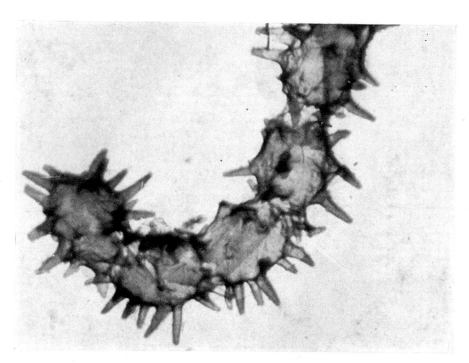

Actinomyces (Streptomyces) coerulescens Gause et Preobrazhenskaya 1957, INA 1581

Spore surface: spiny intermediate long
Mineral medium No. 1
Carbon replica

Figure 35
20 000 ×

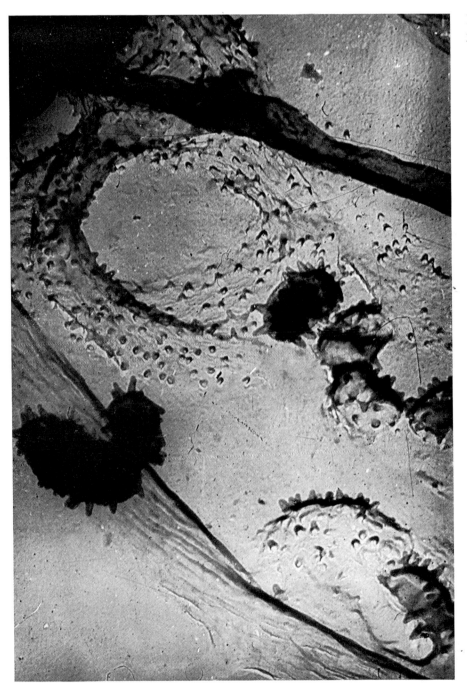

Streptomyces (Actinomyces) purpurascens Lindenbein 1952 NRRL 1454

Spore surface: spiny intermediate long
Mineral medium No. 1

Figure 36
16 000 ×

Figure 37
40 000 ×

Streptomyces (Actinomyces) hirsutus Ettlinger et al. 1958
ISP 5095

Spore surface: spiny intermediate long
Mineral medium No. 1

Figure 38
10 000 ×

Figure 39
40 000 ×

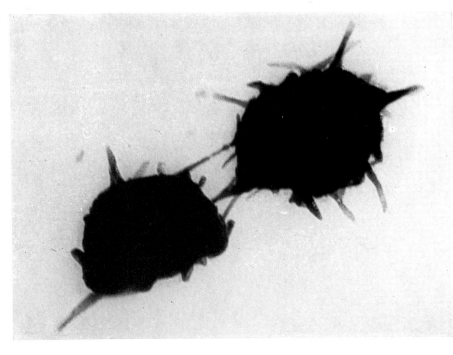

Actinomyces (Streptomyces) janthinus Artamonova et Krassilnikov 1960, INMI 117

Spore surface: spiny intermediate long
Mineral medium No. 1

Figure 40
16 000 ×

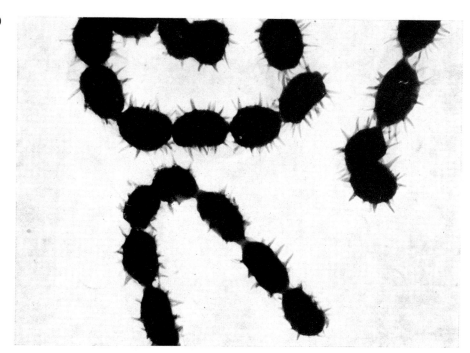

Figure 41
40 000 ×

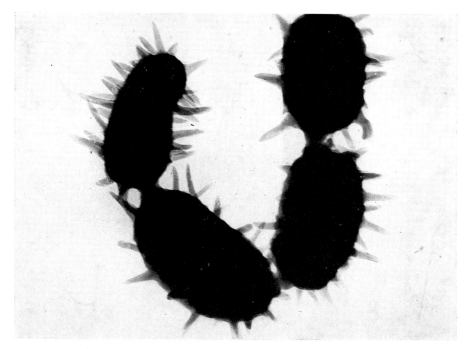

Streptomyces (Actinomyces) echinatus Corbaz et al. 1957
ETH 8331

Spore surface: spiny intermediate long
Oatmeal medium

Figure 42
20 000 ×

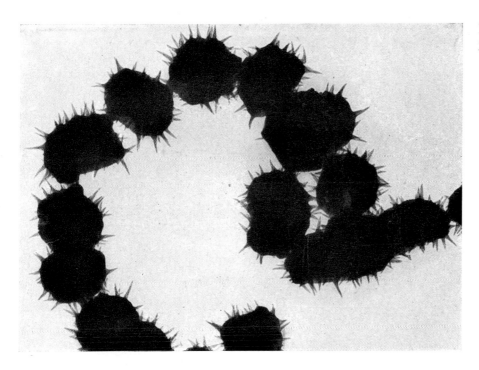

Figure 43
40 000 ×

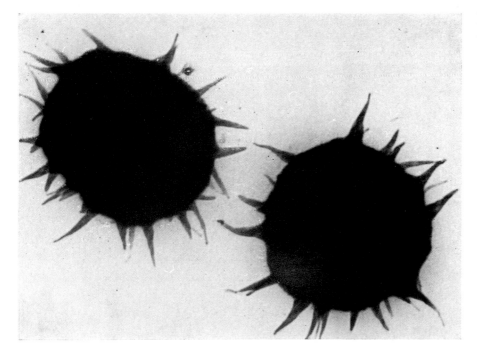

E

Streptomyces (Actinomyces) chartreusis Calhoun et Johnson 1956, NRRL 2287

Spore surface: spiny intermediate long
Mineral medium No. 1

Figure 44
16 000 ×

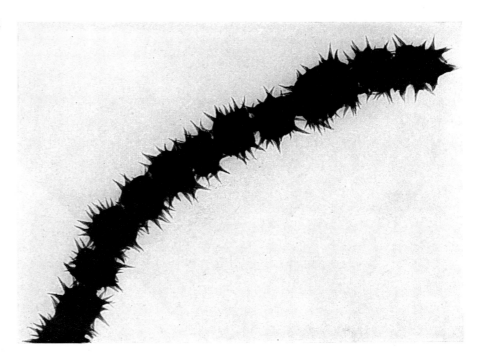

Figure 45
40 000 ×

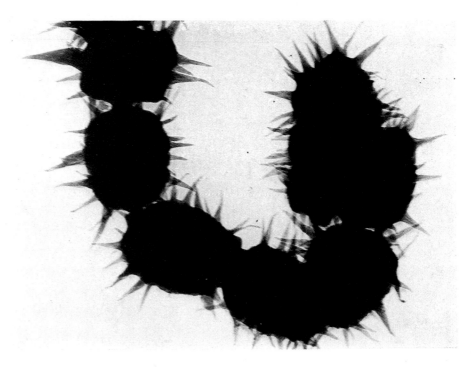

Actinomyces (Streptomyces) albocyaneus Krassilnikov et Agre 1960, INMI 679

Spore surface: spiny short
Mineral medium No. 1

Figure 46
20 000 ×

Figure 47
40 000 ×

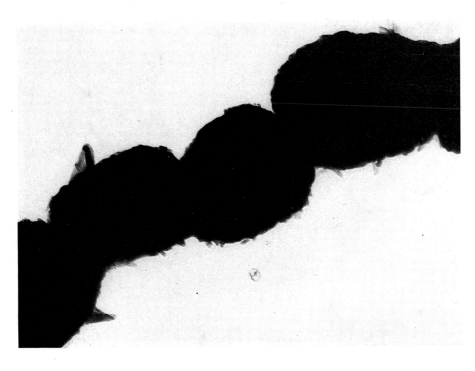

Actinomyces (Streptomyces) sp. Preobrazhenskaya 1960 INA 5092

Spore surface: spiny short
Mineral medium No. 1

Figure 48
16 000 ×

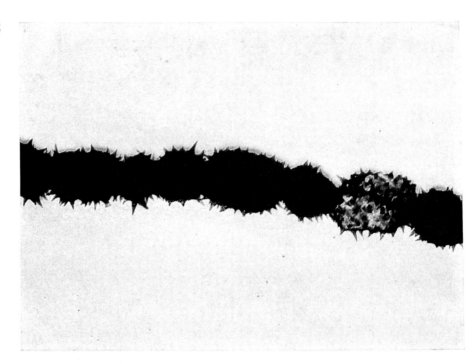

Figure 49
40 000 ×

Actinomyces (Streptomyces) sp. Preobrazhenskaya 1960
INA 5092

Spore surface: spiny short
Mineral medium No. 1
Carbon replica

Figure 50
14 000 ×

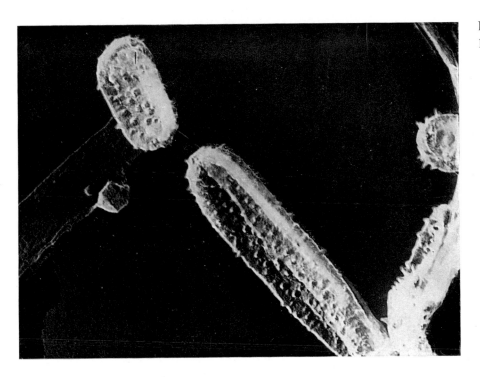

Actinomyces (Streptomyces) rubiginosus Gause et al. 1957 INA 1920

Spore surface: spiny short
Mineral medium No. 1

Figure 51
10 000 ×

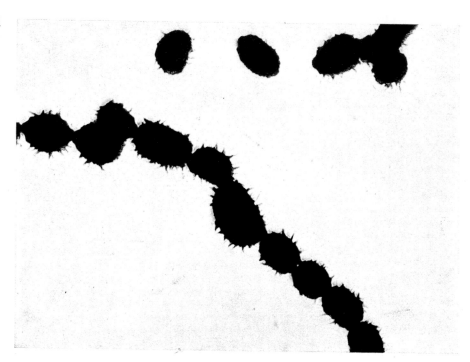

Figure 52
40 000 ×

Streptomyces (Actinomyces) filipinensis Ammann et al. 1955
ISP 5112

Spore surface: spiny short
Synthetic medium No. 1

Figure 53
10 000 ×

Figure 54
40 000 ×

Actinomyces (Streptomyces) griseorubens Gause et al. 1957
INA 10239

Spore surface: spiny short
Mineral medium No. 1

Figure 55
16 000 ×

Figure 56
40 000 ×

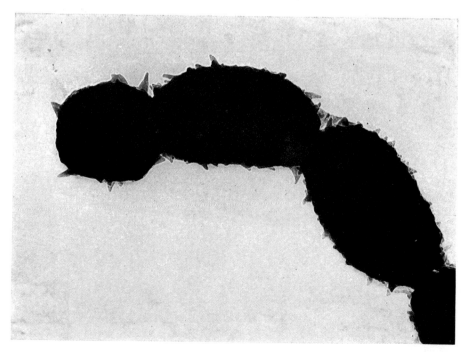

Actinomyces (Streptomyces) sp. Preobrazhenskaya 1957
INA 6508

Spore surface: warty
Mineral medium No. 1

Figure 57
16 000 ×

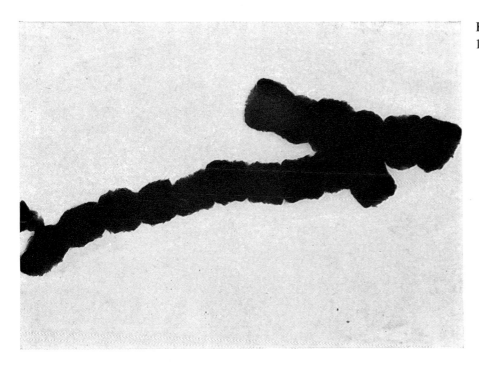

Figure 58
40 000 ×

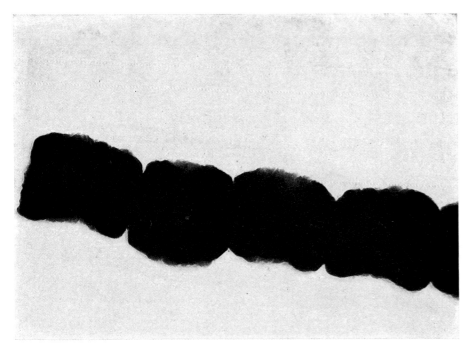

Actinomyces (Streptomyces) sp. Preobrazhenskaya 1957 INA 6508

Spore surface: warty
Mineral medium No. 1
Carbon replica

Figure 59
45 000 ×

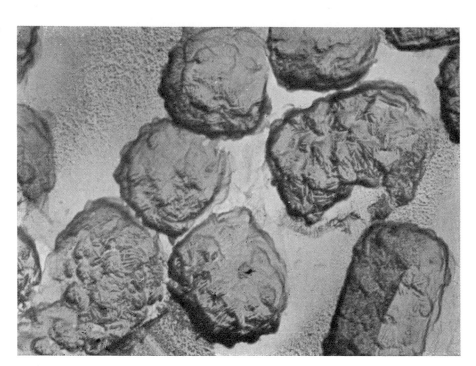

Streptomyces (Actinomyces) luteofluorescens Shinobu ISP 5398

Spore surface: warty
Yeast-extract malt-extract medium

Figure 60
16 000 ×

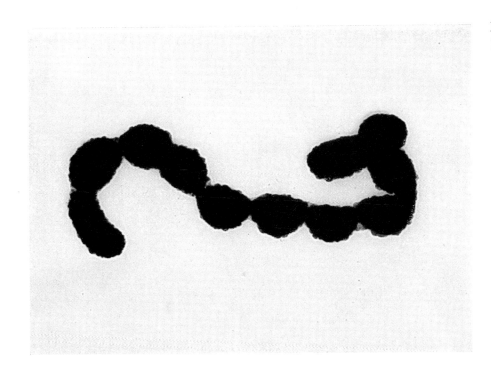

Figure 61
40 000 ×

Actinomyces (Streptomyces) lateritius Gause et al. 1957
INA 6993, ISP 5163

Spore surface: warty
Mineral medium No. 1

Figure 62
16 000 ×

Figure 63
40 000 ×

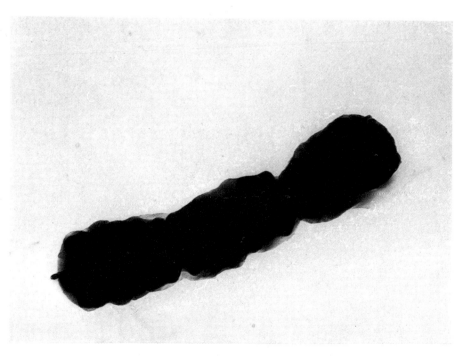

Streptomyces (Actinomyces) lilacinus Nakazawa et al. 1957
4-216

Spore surface: smooth, oval
Mineral medium No. 1

Figure 64
20 000 ×

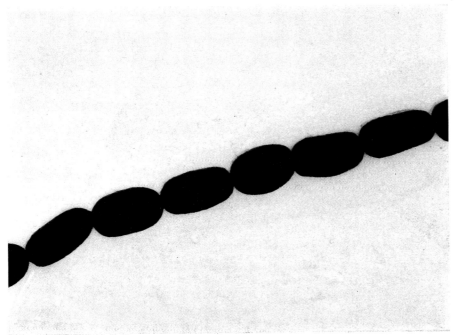

Figure 65
40 000 ×

Actinomyces (Streptomyces) mutabilis Gause et al. 1957 INA 4146

Spore surface: smooth, oval
Mineral medium No. 1

Figure 66
16 000 ×

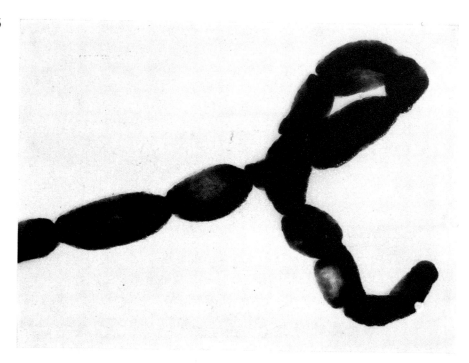

Figure 67
40 000 ×

Actinomyces (Streptomyces) mutabilis Gause et al. 1957
INA 4146

Spore surface: smooth, oval
Mineral medium No. 1
Carbon replica

Figure 68
28 000 ×

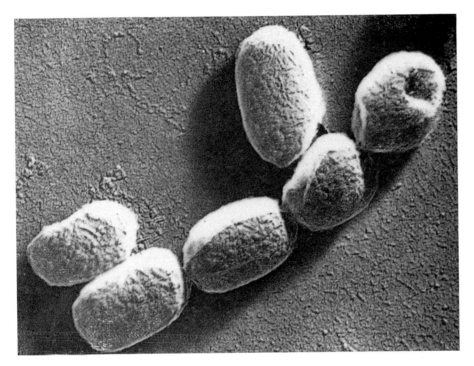

Streptomyces (Actinomyces) caelestis DeBoer et al. 1955 INA 21

Spore surface: smooth, oval
Mineral medium No. 1

Figure 69
16 000 ×

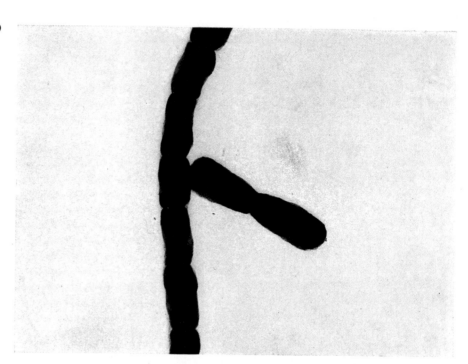

Figure 70
40 000 ×

Streptomyces (Actinomyces) caelestis DeBoer et al. 1955
INA 21

Spore surface: smooth, oval
Mineral medium No. 1
Carbon replica

Figure 71
20 000 $\times$

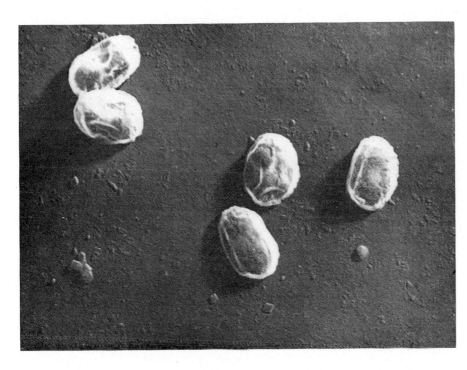

Actinomyces (Streptomyces) globisporus var. flavofuscus Gause et al. 1957, INA 1140

Spore surface: smooth, oblong and oval
Mineral medium No. 1

Figure 72
16 000 ×

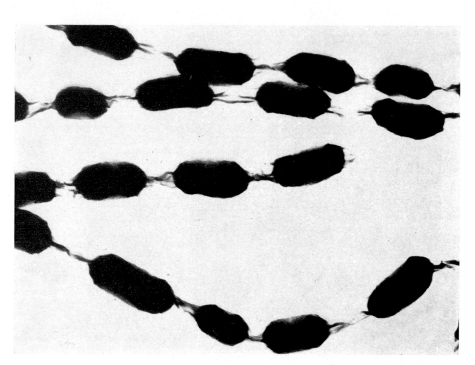

Figure 73
40 000 ×

Streptomyces (Actinomyces) luteolutescens Yen Hsun-chu 1957, 220

Spore surface: smooth, oval, short cylindrical
Mineral medium No. 1

Figure 74
16 000 ×

Figure 75
40 000 ×

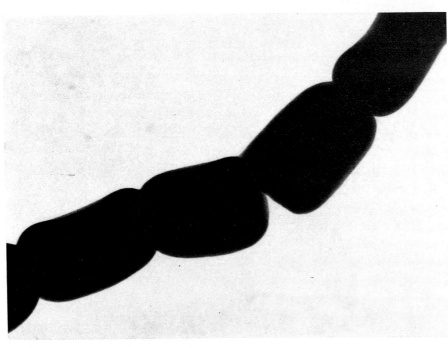

Actinomyces (Streptomyces) cylindrosporus Krassilnikov 1941, INA 2407

Spore surface: smooth, cylindrical short
Mineral medium No. 1

Figure 76
16 000 ×

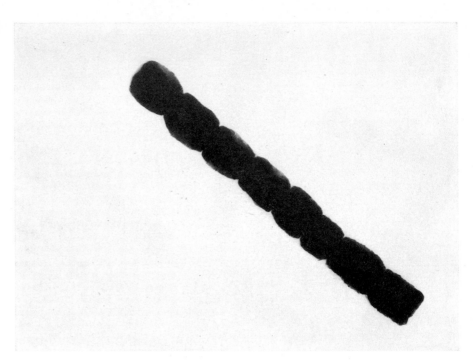

Figure 77
40 000 ×

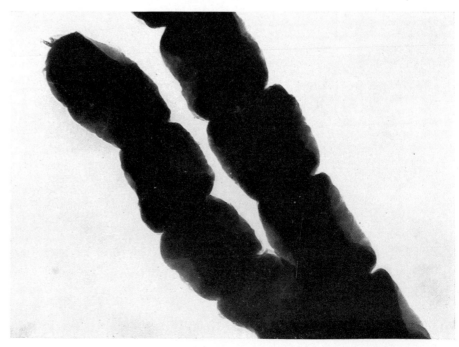

Actinomyces (Streptomyces) cremeus Gause et al. 1957
INA 815/54

Spore surface: smooth, oblong
Mineral medium No. 1

Figure 78
16 000 ×

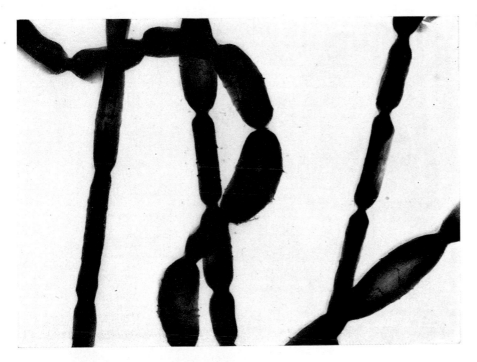

Figure 79
40 000 ×

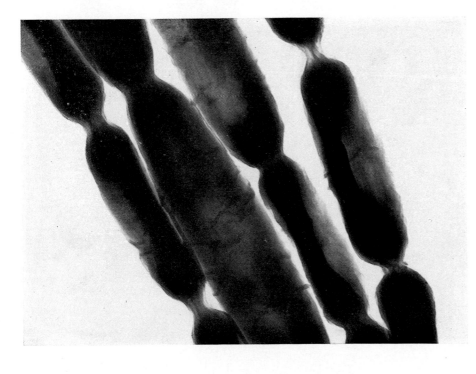

2

Ultrathin sections of Streptomyces (Actinomyces)

Figures 80—87

Actinomyces (Streptomyces) variabilis var. roseolus
Gause et al. 1957, INA 2366

Transversal section through the spores. The central part of the cells with poorly condensed nuclear chromatin is surrounded by more electron dense cytoplasm containing numerous ribosomes. Thick cell membrane exhibiting different electron density preserves spore integrity. From the external, dense layer of the membrane several thin hairy processes are protruding. Uranyl acetate and lead citrate staining.

Mineral medium No. 1

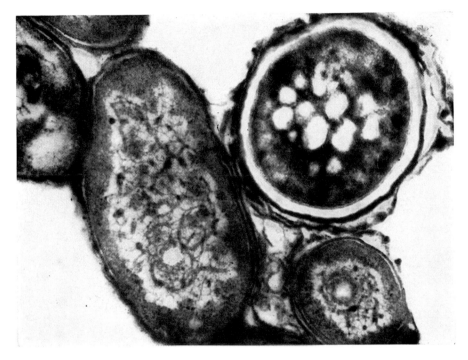

Figure 80
50 000 ×

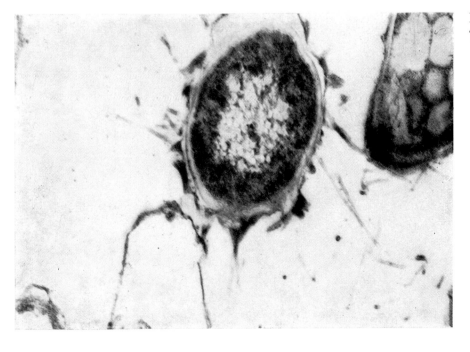

Figure 81
70 000 ×

Streptomyces (Actinomyces) fasciculatus McCormick et al. 1953, ISP 5054

Transversal section through the spores. In the central part of cells threads of poorly condense nuclear chromatin are surrounded by more electron dense cytoplasm filled with numerous ribosomes. Thick cell membrane exhibiting different electron density preserves spore integrity. The spiny processes are protruding from the external more dense layer of cell membrane. The more external silkworm-like membrane represents residual structure of mycelial origin. Uranyl acetate and lead citrate staining.

Mineral medium No. 1

Figure 82
42 000 ×

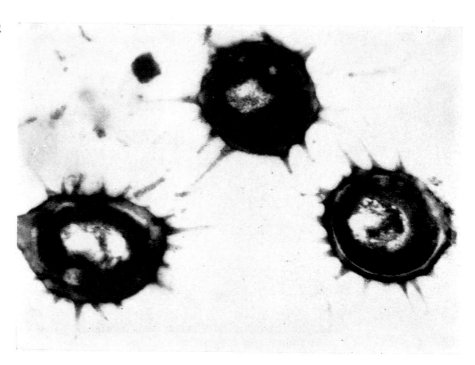

Figure 83
60 000 ×

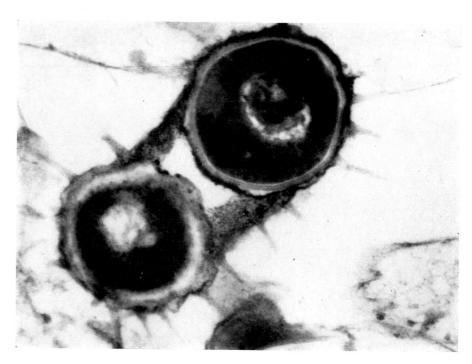

Streptomyces (Actinomyces) luteofluorescens Shinobu ISP 5398

Transversal section through the spores and aerial mycelium. In the central part of spores single threads of nuclear chromatin are surrounded by dense homogeneous cytoplasm. The thick membrane of different electron density preserves spore integrity. Some of spores exhibit double membrane system. The external one represents probably residual membrane structures of mycelial origin. Uranyl acetate and lead citrate staining.

Mineral medium No. 1

Figure 84
40 000 ×

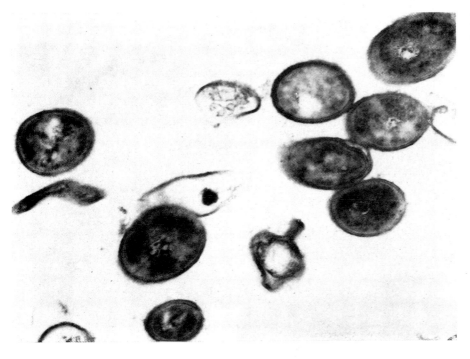

Figure 85
60 000 ×

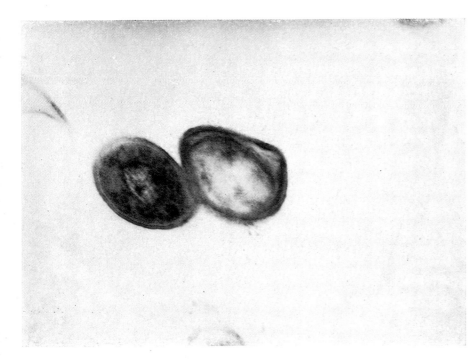

Streptomyces (Actinomyces) niveus Smith et al. 1956
ISP 5088

Transversal section through the spores. The central part of cells with poorly condensed nuclear chromatin is surrounded by more electron dense cytoplasm filled with numerous ribosomes. Thick membrane exhibiting different electron density preserves the spore integrity. The external silkworm-like membrane represents residual structure of mycelial origin. Uranyl acetate and lead citrate staining.

Mineral medium No. 1

Figure 86
43 000 ×

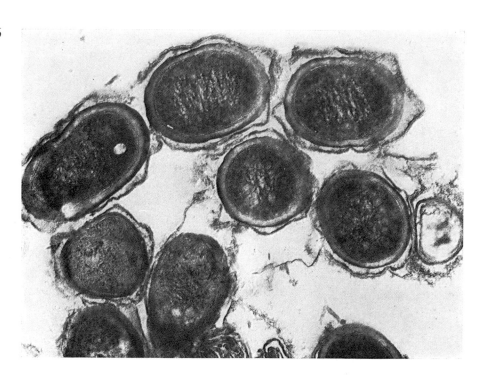

Figure 87
92 000 ×

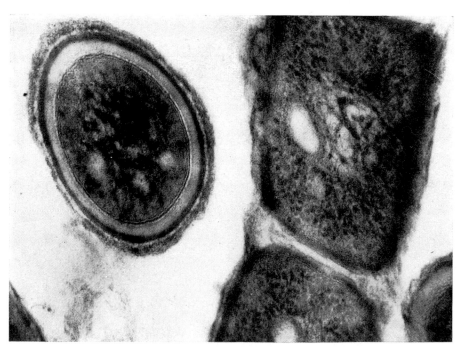

3

Some genera of Streptomycetaceae

Figures 88—108

Actinopycnidium coeruleum Krassilnikov 1962, 729

Pycnidium and smooth spores
Organic medium No. 2

Figure 88
20 000 ×

Figure 89
40 000 ×

Chainia ochracea Thirumalachar 1955, 710

Spore surface: smooth, oval
Organic medium No. 2

Figure 90
16 000 ×

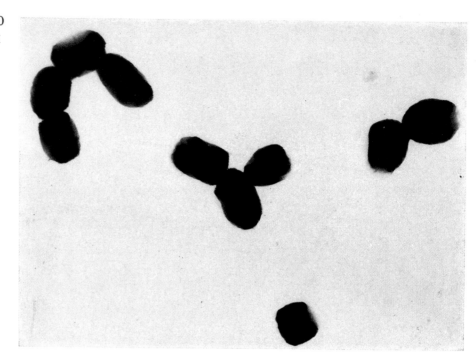

Figure 91
40 000 ×

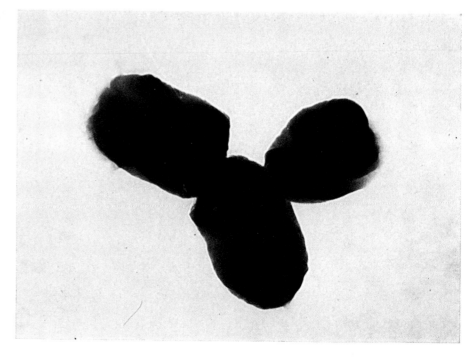

Micromonospora brunnea Sveshnikova, Maximova, Kudrina 1968, INA 166

Spore surface: irregular
Mineral medium No. !

Figure 92
10 000 ×

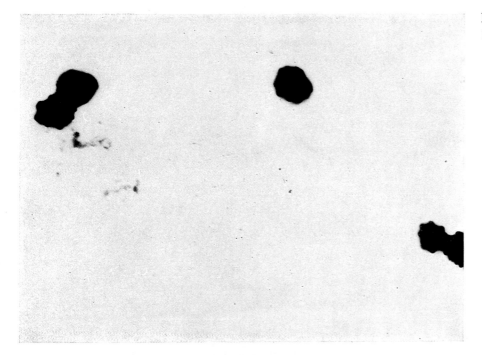

Figure 93
40 000 ×

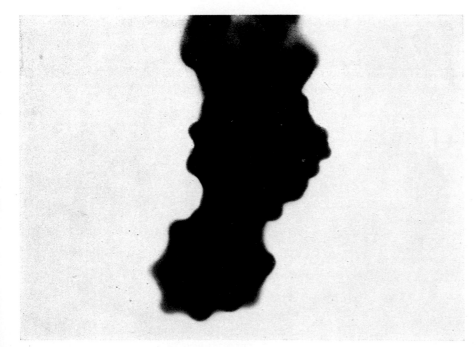

Micromonospora chalcea (Foulerton) Ørskov 1923
ATCC 12452

Spore surface: smooth
Mineral medium No. 1

Figure 94
10 000 ×

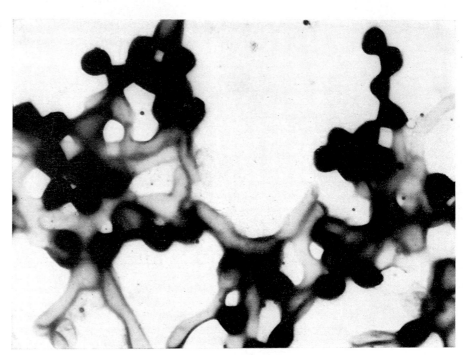

Figure 95
20 000 ×

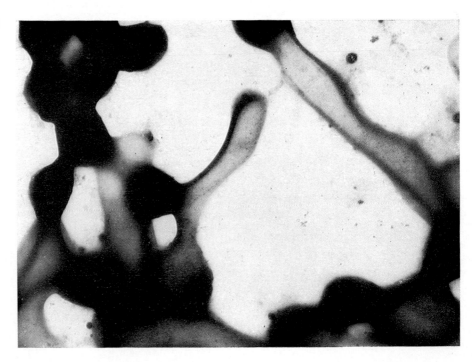

Micromonospora fusca Jensen 1932, strain CBS

Spore surface: smooth
Organic medium No. 2

Figure 96
10 000 ×

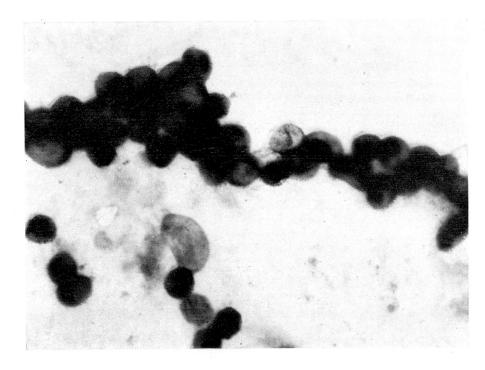

Micromonospora rubra Sveshnikova, Maximova et Kudrina 1968, INA 325

Spore surface: irregular
Mineral medium No. 1

Figure 97
10 000 ×

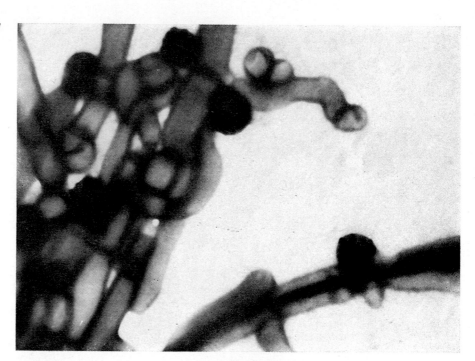

Figure 98
20 000 ×

Micromonospora sp. INA 125

Spore surface: irregular
Mineral medium No. 1

Figure 99
10 000 ×

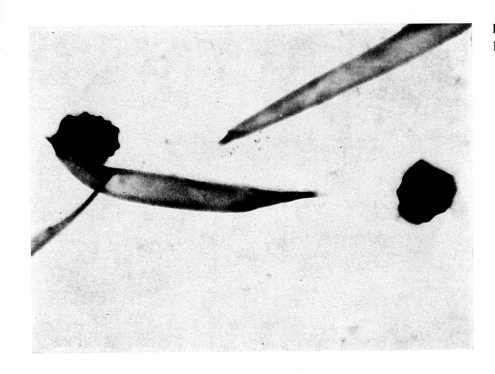

Waksmania Lechevalier et al. 1957, syn. Microbispora parva Nonomura 1960

Bispores
Organic medium No. 2

Figure 100
20 000 ×

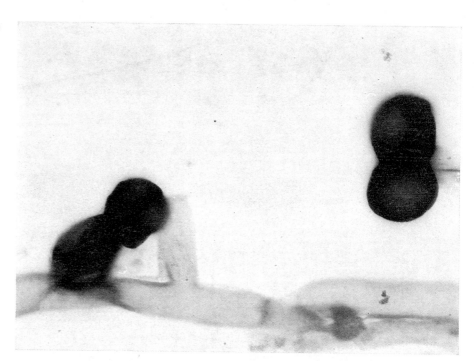

Figure 101
40 000 ×

Actinobifida chromogena Krassilnikov et Agre 1965, 2900

Spore surface: spiny
Organic medium No. 2

Figure 102
10 000 ×

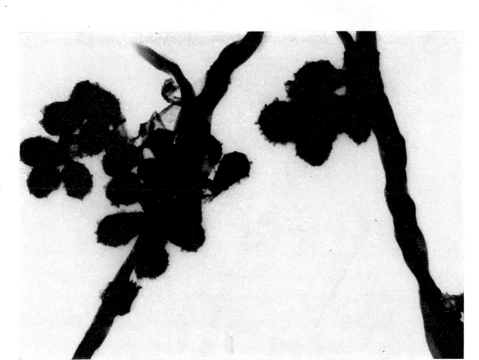

Figure 103
40 000 ×

Thermoactinomyces viridis Shuurmans et al. 1956

Single spores
Organic medium No. 2

Figure 104
10 000 ×

Figure 105
20 000 ×

Microellobosporia (Macrospora) violacea Tsyganov et al. 1963
LIA 2732

Sporangium
Organic medium No. 2

Figure 106
16 000 ×

Amorphosporangium auranticolor Couch 1963, 819

Spores in pseudosporangium
Organic medium No. 2

Figure 107
10 000 ×

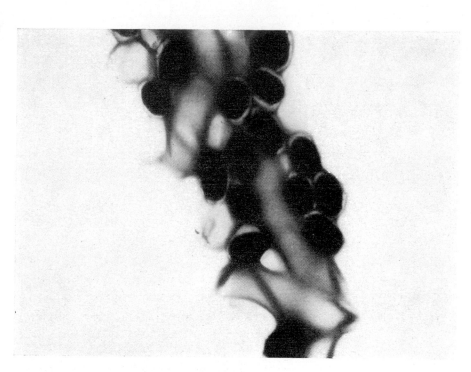

Figure 108
20 000 ×

4

Ultrathin sections of some genera of Streptomycetaceae

Figures 109—124

Actinopycnidium coeruleum Krassilnikov 1962, 729

Transversal section through the pycnidium. Note the well developed single spore in the pycnidium sac. No ultrastructure differences are seen between the pycnidium sac wall and cell membrane of spore. Uranyl acetate and lead citrate staining.

Organic medium No. 2

Figure 109
80 000 ×

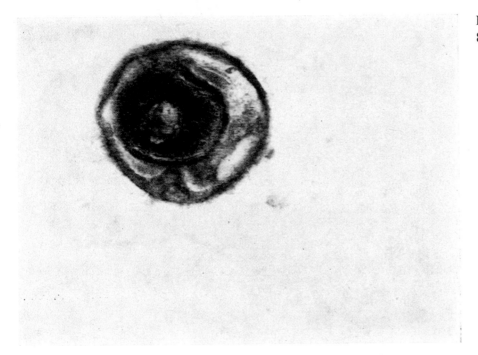

Figure 110
100 000 ×

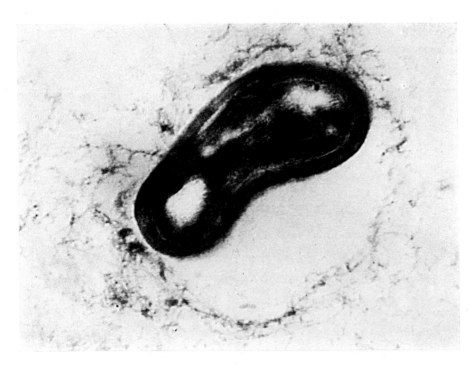

Chainia ochracea Thirumalachar 1955, 710

Transversal section through the chain of spores. In the dense cytoplasm of spores no nucleus is seen. The thick cell membrane preserves spore integrity. The desmosome-like bridges between spores connect them to form the chain. Uranyl acetate and lead citrate staining.

Organic medium No. 2

Figure 111
35 000 ×

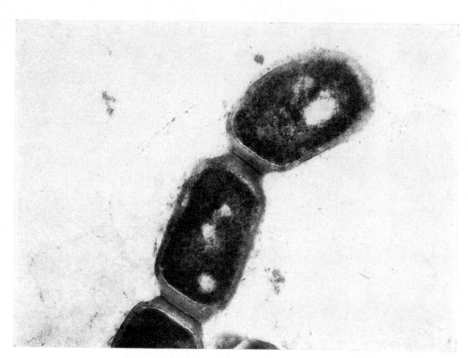

Figure 112
70 000 ×

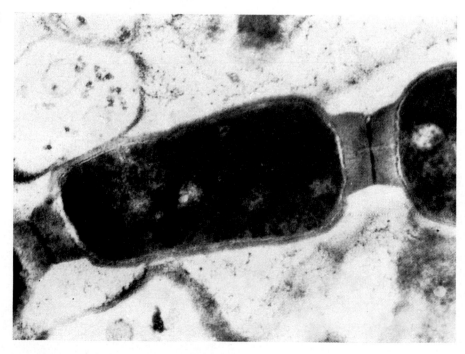

Micromonospora chalcea Foulerton Ørskov 1923
ATCC 12452

Transversal section through the spores and aerial mycelium. In the central part of the spores packed threads of nuclear chromatin surrounded by more electron dense cytoplasm filled with numerous ribosomes. Thick membrane of different electron density preserves spore integrity. Uranyl acetate and lead citrate staining.

Mineral medium No. 1

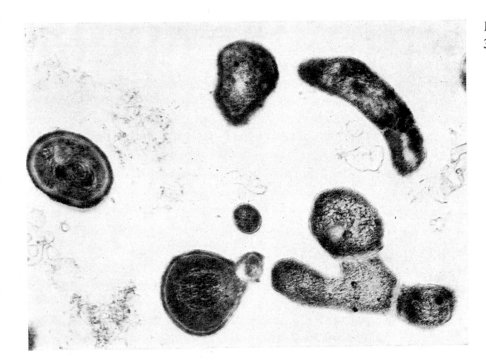

Figure 113
30 000 ×

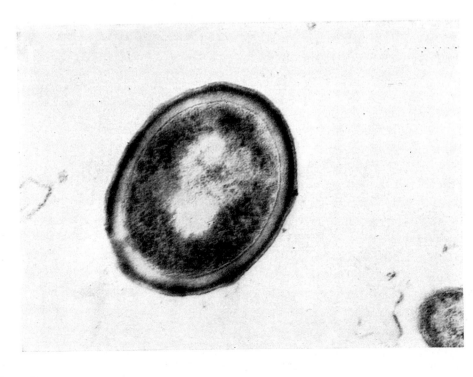

Figure 114
60 000 ×

Waksmania Lechevalier et al. 1957, syn. Microbispora parva Nonomura 1960

Transversal and longitudinal section through aerial mycelium and spores. The characteristic pair of spores each with independent nucleus and cytoplasmic structures separated by cell membrane. The thick cell membrane exhibiting different electron density preserves spore integrity. Uranyl acetate and lead citrate staining.

Organic medium No. 2

Figure 115
23 000 ×

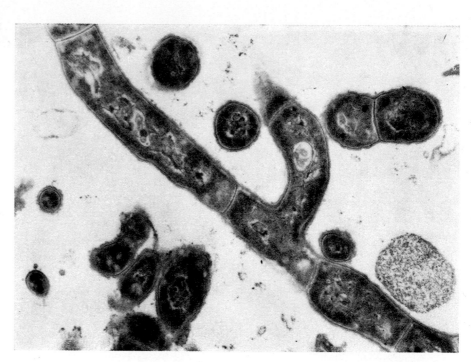

Figure 116
43 000 ×

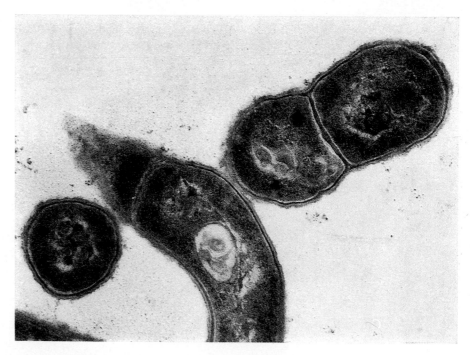

Transversal section through the chain of spores. In the central part of cells poorly condensed nuclear chromatin is surrounded by more electron dense cytoplasm filled with numerous ribosomes. The thick cell membrane exhibiting different electron density preserves spore integrity. On the surface of the spores folding of cell membrane forms short spiny-like protrusions. Uranyl acetate and lead citrate staining.

Organic medium No. 2

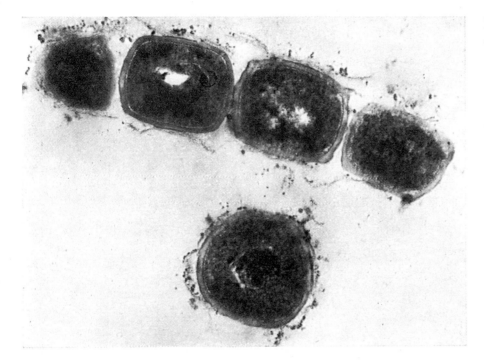

Figure 117
40 000 ×

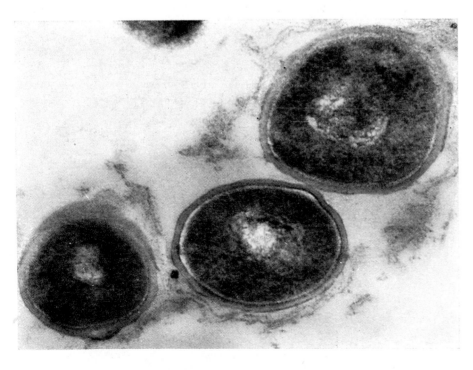

Figure 118
50 000 ×

Thermoactinomyces viridis Shuurmans et al. 1956

Transversal and longitudinal sections through aerial mycelium. Note the variety in ultrastructure organization depending on functional activity of different parts of mycelium. Uranyl acetate and lead citrate staining.

Organic medium No. 2

Figure 119
16 000 ×

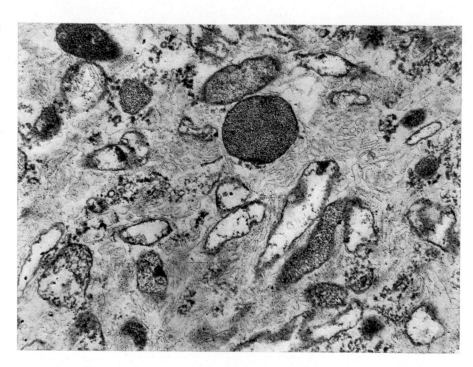

Figure 120
30 000 ×

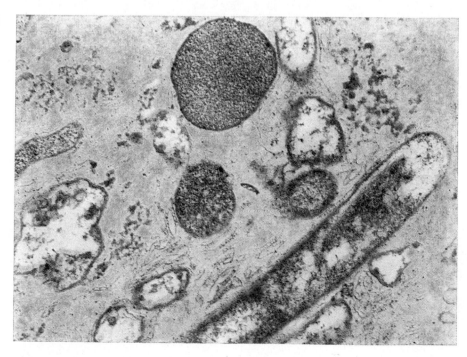

Actinosporangium violaceum Krassilnikov et al. 1961, 655

Transversal section through the aerial mycelium and spores. The inner part of mycelium is filled with vesicular structures containing homogenous material of low electron density. Spores surrounded by the thick membrane exhibiting different electron density contain nucleus and cytoplasm filled with ribosomes. The electron dense round bodies dispersed throughout the inner part of mycelium are most probably lipid droplets. Uranyl acetate and lead citrate staining.

Organic medium No. 2

Figure 121
38 000 ×

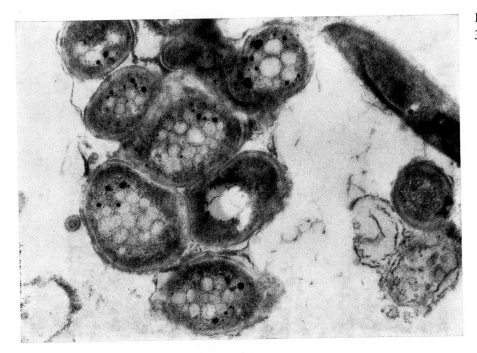

Figure 122
62 000 ×

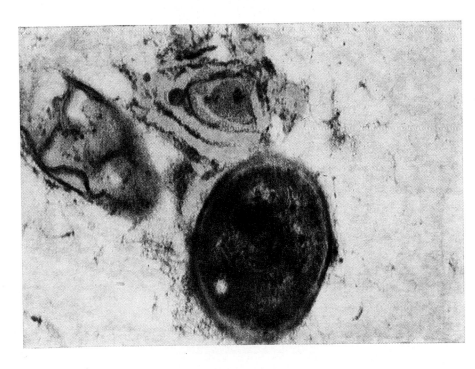

Amorphosporangium auranticolor Couch 1963, 819

Transversal and longitudinal section through the spores and aerial mycelium. The central part of spores with poorly condensed nuclear chromatin is surrounded by more electron dense cytoplasm filled with numerous ribosomes. The thick membrane exhibiting different electron density preserves spore integrity. Longitudinal sections of aerial mycelium demonstrate variety of ultrastructure organization depending on its functional differentiation. Uranyl acetate and lead citrate staining.

Organic medium No. 2

Figure 123
40 000 ×

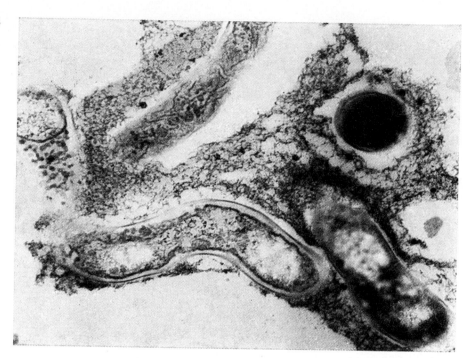

Figure 124
70 000 ×

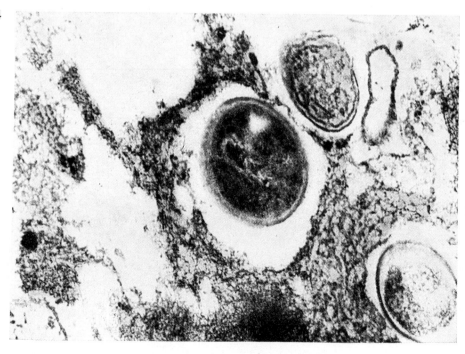

5

Scanning electron microscopy

Figures 125—137

Streptomyces finlayi Szabo et al. 1963, ISP 5218

Spore surface: hairy and smooth

Figure 125
10 000 ×

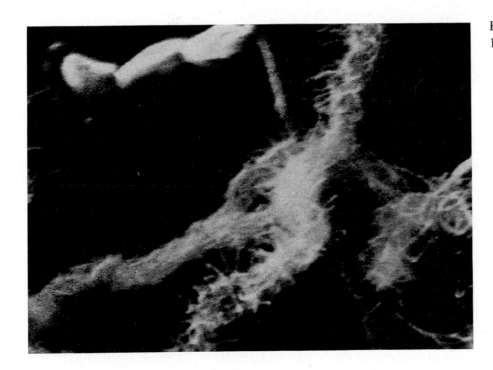

Streptomyces scabies (Taxter 1891) Waksman et Henrici 1948 CBS

Spore surface: spiny (on surface of potato)

Figure 126
5 500 ×

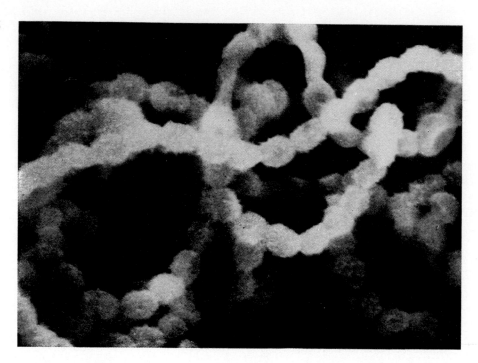

Streptomyces viridosporus Parke et al. 1954, ISP 5243

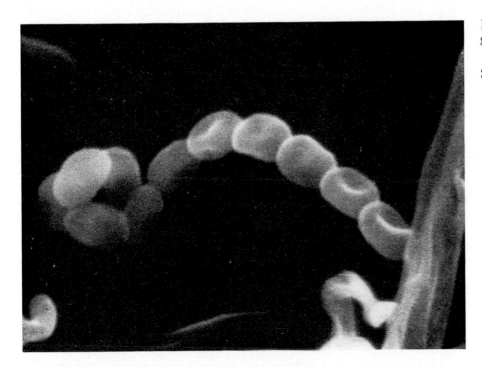

Figure 127
8 000 ×

Spore surface: smooth

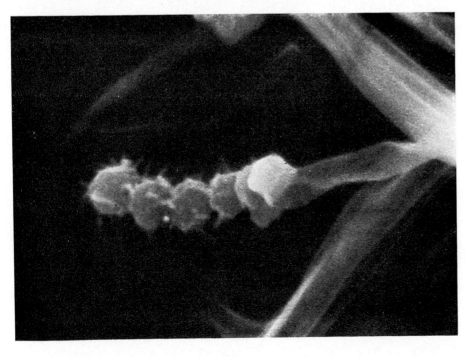

Figure 128
8 000 ×

Spore surface: spiny

Streptomyces viridosporus Parke et al. 1954, ISP 5243

Spore surface: smooth, spiny

Figure 129
8 000 ×

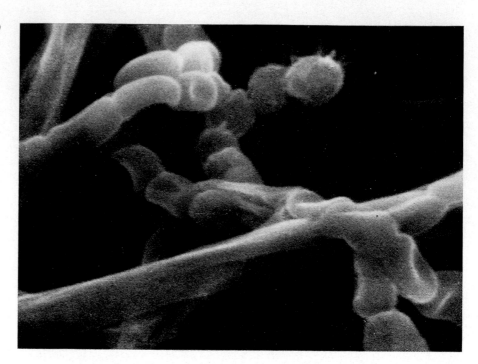

Actinopycnidium coeruleum Krassilnikov 1962, RIA 729

Spore surface: smooth

Figure 130
8 000 ×

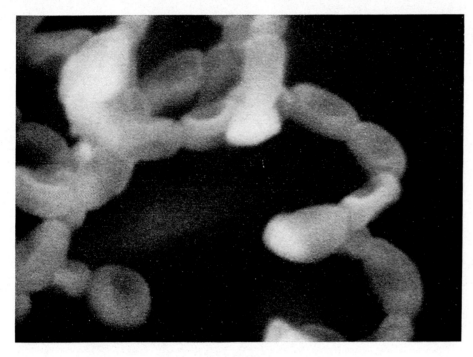

Micromonospora melanosporea (Krainsky 1914) Baldacci et Locci 1961, Baldacci 583

Spore surface: smooth

Figure 131
8 000 ×

Thermomonospora viridis (Shuurmans et al.) Küster et Locci 1963, Cross 78

Spore surface: smooth

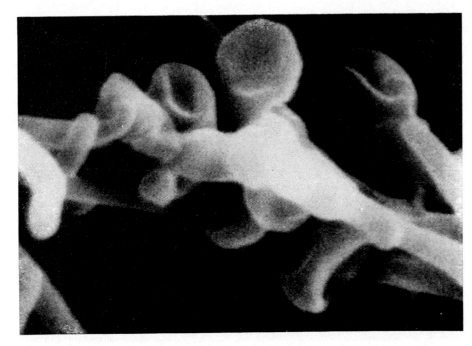

Figure 132
10 000 ×

Actinobifida dichotomica Krassilnikov et Agre 1964, Cross 339

Spore surface: smooth, polygonal shaped spores

Figure 133
5 000 ×

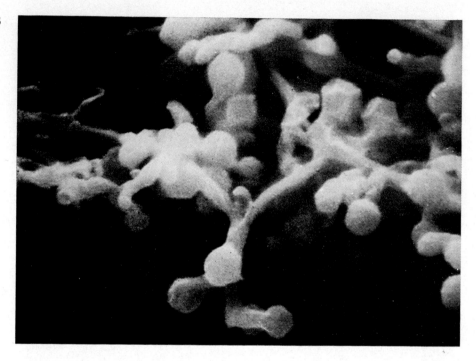

Thermoactinomyces vulgaris Tsiklinsky 1899, Küster P 121

Spore surface: smooth, polygonal shaped spores

Figure 134
8 000 ×

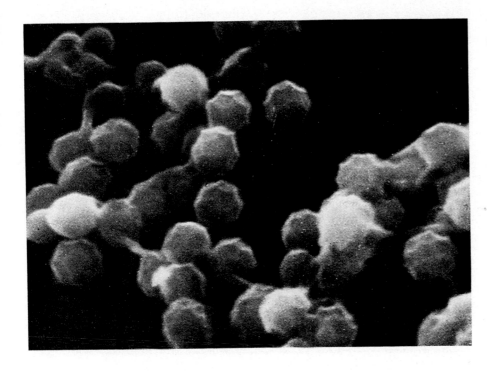

Microellobosporia flavea Cross et al. 1963, IMRU 3858

Young sporangium

Figure 135
8 000 ×

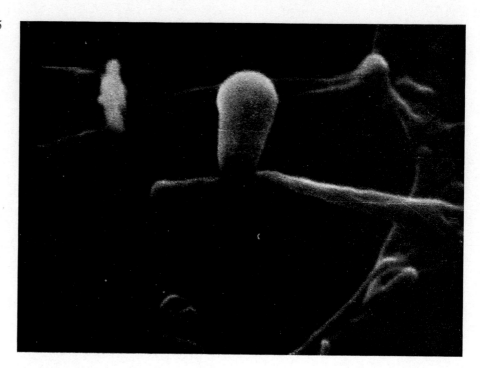

Streptosporangium sp.

Single sporangium

Figure 136
8000 ×

Actinoplanes sp.

Surface of mature sporangium

Figure 137
8 000×

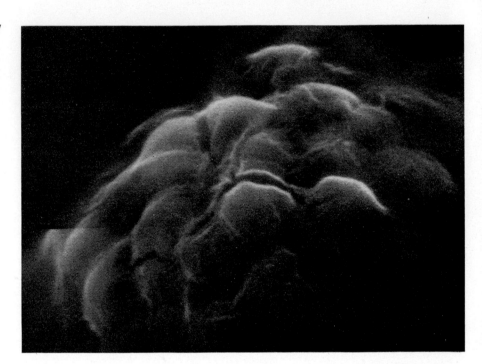

Appendix

List of strains seen in electron microscope tabulated according to spore surface morphology

Spore surface: hairy

Actinomyces (Streptomyces) acrimycini Gause et al. 1957, INA 7699
Actinomyces (Streptomyces) atrovirens Maximova 1968, INA 300
Streptomyces (Actinomyces) finlayi Szabo et al. 1963, ISP 5218
Streptomyces (Actinomyces) flaveolus (Waksman), Waksman et Henrici 1948, ATCC 3319
Actinomyces (Streptomyces) flavochromogenes var. heliomycini Gause et al. 1958, INA 2915
Actinomyces (Streptomyces) flavoviridis Krassilnikov 1941, INA 2314/53
Actinomyces (Streptomyces) glaucescens Gause et Preobrazhenskaya 1957, INA 13386
Actinomyces (Streptomyces) glaucescens Gause et Preobrazhenskaya 1957, INA 13380
Actinomyces (Streptomyces) griseomycini Gause et al. 1957, INA 3777
Actinomyces (Streptomyces) griseus Krainsky 1914, Krassilnikov 1941, emend. Preobrazhenskaya 1966, INA 605
Actinomyces (Streptomyces) malachiticus Gause et al. 1957, INA 7035
Actinomyces (Streptomyces) nordicus Preobrazhenskaya 1966, INA 7551
Streptomyces (Actinomyces) pilosus Ettlinger et al. 1958, ISP 5097
Streptomyces (Actinomyces) prasinopilosus Ettlinger et al. 1958, ISP 5098
Actinomyces (Streptomyces) prasinoviolaceus sp. nov. Maximova 1955, INA 352
Actinomyces (Streptomyces) variabilis var. roseolus Gause et al. 1957, INA 8820/54
Actinomyces (Streptomyces) viridiviolaceus Gause et al. 1957, INA 313

Spore surface: spiny

Streptomyces (Actinomyces) afghaniensis Shimo et al. 1959, ISP 5228
Actinomyces (Streptomyces) albocyaneus Krassilnikov et Agre 1960, INMI 679
Actinomyces (Streptomyces) atroolivaceus var. mutomycini Gause et al. 1959, INA 4305
Actinomyces (Streptomyces) bicolor Gause et Preobrazhenskaya 1957, INA 5104
Streptomyces (Actinomyces) canus Heinemann et al. 1953, ATCC 12237
Streptomyces (Actinomyces) cellostaticus Tohoku 1958, ISP 5189
Streptomyces (Actinomyces) chartreusis Calhoun et al. 1956, NRRL 2287
Actinomyces (Streptomyces) chlorinus Sveshnikova 1968, INA 470
Actinomyces (Streptomyces) chlorobiensis Krassilnikov et al. 1965, INMI 6166
Actinomyces (Streptomyces) chromogenes Lachner-Sandoval, emend. Krassilnikov 1941, INA 6085
Actinomyces (Streptomyces) coeruleofuscus Gause et Preobrazhenskaya 1957, INA 6920
Actinomyces (Streptomyces) coeruleofuscus var. actinomycini Maximova et al. 1954, INA 2206
Actinomyces (Streptomyces) coeruleoroseus Preobrazhenskaya 1966, INA 9106
Actinomyces (Streptomyces) coeruleorubidus Gause et Preobrazhenskaya 1957, INA 12531/54
Actinomyces (Streptomyces) coeruleorubidus Gause et al. 1966, INA 11654
Actinomyces (Streptomyces) coerulescens Gause et Preobrazhenskaya 1957, INA 4562
Streptomyces (Actinomyces) curacoi Cataldi et al. 1962, ATCC 13385
Actinomyces (Streptomyces) cyanoalbus Krassilnikov et Agre 1960, INMI 435

Streptomyces (Actinomyces) echinatus Corbaz et al. 1957, ETH 8331
Streptomyces (Actinomyces) fasciculatus McCormick et al. 1953, ISP 5054
Streptomyces (Actinomyces) filipinensis Ammann et al. 1955, ISP 5112
Actinomyces (Streptomyces) griseoflavus Krainsky 1914, Waksman et Henrici 1948, ISP 5456
Actinomyces (Streptomyces) griseorubens Gause et al. 1957, INA 10239
Streptomyces (Actinomyces) hirsutus Ettlinger et al. 1958, ISP 5095
Streptomyces (Actinomyces) ipomoea Person et Martin 1940, ATCC 11747, INA 1175
Actinomyces (Streptomyces) incarnatus Sveshnikova 1966, INA 13866
Actinomyces (Streptomyces) janthinus Artamonova et Krassilnikov 1960, INMI 117
Actinomyces (Streptomyces) longoechinatus Maximova 1968, INA 2722
Actinomyces (Streptomyces) lucinus Preobrazhenskaya 1968, INA 9997
Streptomyces (Actinomyces) malachiticus Gause et al. 1957, ISP 5167
Actinomyces (Streptomyces) malachitofuscus Preobrazhenskaya et al. 1964, INA
Actinomyces (Streptomyces) mellinus Maximova et al. 1965, INA 2969
Streptomyces (Actinomyces) noursei Hazen et Brown 1950, NCIB 8593
Streptomyces (Actinomyces) prasinisporus Tresner et al. 1966, ATCC 17918
Streptomyces (Actinomyces) purpurascens Lindenbein 1952, NRRL 1454
Actinomyces (Streptomyces) rubiginosus Gause et al. 1957, INA 1920
Actinomyces (Streptomyces) spinosus Preobrazhenskaya 1966, INA 3763
Actinomyces (Streptomyces) subcaucasicus Sveshnikova et al. 1966, INA 9040
Streptomyces (Actinomyces) thermotolerans Pagano 1959, ISP 5227
Streptomyces (Actinomyces) toyokaensis Nishimura 1956, holotype 112
Actinomyces (Streptomyces) valynus Preobrazhenskaya 1966, INA 612
Actinomyces (Streptomyces) variabilis Gause et al. 1957, INA 5557/54
Actinomyces (Streptomyces) violaceoniger var. crystallomycini Gause et al. 1957, INA 5148
Actinomyces (Streptomyces) viridichromogenes Krainsky 1914, emend. Krassilnikov 1949, ATCC 3356
Streptomyces (Actinomyces) viridosporus Brit. Pat. 1954, ISP 5243
Actinomyces (Streptomyces) sp. Preobrazhenskaya 1960, INA 5092

Spore surface: warty

Actinomyces (Streptomyces) lateritius Gause et al. 1957, INA 6993
Actinomyces (Streptomyces) viridans Krassilnikov 1941, INMI 1671
Actinomyces (Streptomyces) sp. Preobrazhenskaya 1957, INA 6508
Actinomyces (Streptomyces) sp. Preobrazhenskaya 1954, INA 8618
Streptomyces (Actinomyces) luteofluorescens Shinobu, ISP 5398

Spore surface: smooth

Streptomyces (Actinomyces) abikoensis Umezawa et al. 1956, NIHJ Z-6-1
Actinomyces (Streptomyces) abikoensum var. spiralis Gause et al. 1957, INA 6355
Streptomyces (Actinomyces) aburaviensis Nishimura et al. 1957, ISP 5033
Streptomyces (Actinomyces) actuosus Pinnert et al. 1961, ISP 5337
Streptomyces (Actinomyces) albidoflavus Rossi-Doria 1891, Waksman et Henrici 1948, ME-1-2, Baldacci
Streptomyces (Actinomyces) albidus Duché 1934, emend. Krassilnikov 1941, INA 11138/54
Streptomyces (Actinomyces) albireticuli Nakazawa 1955, NRRL B-1670
Streptomyces (Actinomyces) albochromogenes Tanaka et al. 1958, holotype
Streptomyces (Actinomyces) albofaciens Thirumalachar 1960, ISP 5268
Streptomyces (Actinomyces) alboflavus Waksman et Curtis 1916, Waksman et Henrici 1948, ISP 5045
Actinomyces (Streptomyces) albohelveticus Krassilnikov et al. 1965, INMI 1349
Streptomyces (Actinomyces) albolongus Tsukira et al. 1964, 307 K7

Streptomyces (Actinomyces) alboniger Hesseltine et al. 1954, NRRL 1832, ATCC 12461
Streptomyces (Actinomyces) albosporeus Krainsky 1914, Waksman et Henrici 1948, ATCC 3003
Actinomyces (Streptomyces) albovinaceus Gause et al. 1957, INA 273
Streptomyces (Actinomyces) alboviridis Duché 1954, emend. Gause et al. 1957, INA 2605
Streptomyces (Actinomyces) albus Rossi-Doria, emend. Gasperini 1892, Waksman et Henrici 1948, ATCC 3004
Streptomyces (Actinomyces) althioticus Yamaguchi et al. 1957, ISP 5092
Streptomyces (Actinomyces) amacusaensis Nagatsu et al. 1963, INA 7018
Streptomyces (Actinomyces) aminophilus Oswald et al. 1955, ISP 5186
Streptomyces (Actinomyces) annulatus Beijerinck 1912, emend. Krassilnikov 1941, IMRU 3307
Streptomyces (Actinomyces) antibioticus Waksman et Woodruff 1941, Waksman et Henrici 1948, ATCC 10382
Streptomyces (Actinomyces) antimycoticus Waksman et al. 1952, ISP 5284
Actinomyces (Streptomyces) antocyaneus Novogrudsky 1950, INA 1016, holotype
Actinomyces (Streptomyces) arenicolor Maximova 1964, INA 141
Streptomyces (Actinomyces) argenteolus Perlman 1957, ATCC 11009
Actinomyces (Streptomyces) ashchabadicus Preobrazhenskaya 1966, INA 13496
Actinomyces (Streptomyces) atroolivaceus Gause et al. 1957, INA 1580/53
Actinomyces (Streptomyces) aurantiogriseus Gause et al. 1957, INA 10061
Streptomyces (Actinomyces) aureofaciens Duggar 1948, ATCC 10762
Actinomyces (Streptomyces) aureofaciens var. oxytetracyclini Preobrazhenskaya et al. 1961, INA 255
Actinomyces (Streptomyces) aureofasciculus Krassilnikov et al. 1965, ISP 5414
Actinomyces (Streptomyces) aureomonopodeales Krassilnikov et Yuan Chi-shen 1965, ISP 5416
Streptomyces (Actinomyces) aureus Waksman et Curtis 1916, Waksman et Henrici 1948, ATCC 3309
Actinomyces (Streptomyces) aurigineus Krassilnikov et al. 1965, INMI 2375
Actinomyces (Streptomyces) aurini Gause et al. 1957, INA 6001
Streptomyces (Actinomyces) autotrophicus Takamiya et Tubaki 1956, NRRL B-1826
Streptomyces (Actinomyces) azureus Kelly et al. 1959, ATCC 14921
Streptomyces (Actinomyces) baarnensis Duché 1934, ISP 5232
Actinomyces (Streptomyces) bacillaris Nikitina et al. 1960, INMI 445
Actinomyces (Streptomyces) badius Gause et al. 1957, INA 1203/53
Streptomyces (Actinomyces) bikiniensis Johnstone, Waksman 1948, ATCC 11062
Streptomyces (Actinomyces) bottropensis Konink. Nederl. Gist. et Spirit. 1956, ISP 5262
Streptomyces (Actinomyces) canescus Huckey et al. 1952, ISP 5001
Streptomyces (Actinomyces) cacaoi Waksman 1932, Waksman et Henrici 1948, ΛTCC 3082
Streptomyces (Actinomyces) caelestis DeBoer et al. 1955, NRRL 2418, INA 21
Streptomyces (Actinomyces) calvus Backus et al. 1957, PSA 152 T-3018
Streptomyces (Actinomyces) catenulae Dadisson et Finlay 1959, ISP 5258
Streptomyces (Actinomyces) capreolus Stark et al. 1962, NRRL 2773
Actinomyces (Streptomyces) candidus Krassilnikov 1941, INA 5855/55
Actinomyces (Streptomyces) candidus var. alboroseus Gause et al. 1957, INA 4143/54
Actinomyces (Streptomyces) carpaticus Maximova 1966, INA 8851
Streptomyces (Actinomyces) cellulosae Krainsky 1914, Waksman et Henrici 1948, IAUR 2515
Actinomyces (Streptomyces) chromofuscus Gause et al. 1957, INA 6050
Streptomyces (Actinomyces) chrysomallus Lindenbein 1952, Waksman 3657, ATCC 11523
Actinomyces (Streptomyces) chinensis Sveshnikova 1966, INA 6901
Actinomyces (Streptomyces) circulatus var. monomycini Gause et al. 1960, INA 1465
Actinomyces (Streptomyces) cinnabarinus Gause et al. 1957, INA 5933
Streptomyces (Actinomyces) cinnamomeus f. cinnamomeus Benedict et al. 1954, ISP 5005
Streptomyces (Actinomyces) citreus Krainsky 1914, Waksman et Henrici 1948, ATCC 10974
Actinomyces (Streptomyces) citreofluorescens Koreniako et al. 1960, INMI, holotype
Actinomyces (Streptomyces) coelicolor var. flavus Gause et al. 1957, INA 28
Streptomyces (Actinomyces) coelicolor Müller 1908, Krassilnikov 1941, ATCC 3355
Streptomyces (Actinomyces) caelestis DeBoer et al. 1955, NRRL 2418

Streptomyces (Actinomyces) cretaceus Krüger emend. Krassilnikov 1941, INMI 38

Actinomyces (Streptomyces) cylindrosporus Krassilnikov 1941, INA 2407

Actinomyces (Streptomyces) cyaneofuscatus Gause et al. 1957, INA 99/54

Streptomyces (Actinomyces) diastatochromogenes Krainsky 1914, Waksman et Henrici 1948, NRRL 778

Streptomyces (Actinomyces) diastaticus Krainsky 1914, Waksman et Henrici 1948, INA 8011

Streptomyces (Actinomyces) echimensis Nakazawa 1954, ISP 5233

Streptomyces (Actinomyces) endus Gottlieb et Carter 1956, ISP 5187

Streptomyces (Actinomyces) erythreus Waksman et Curtis 1916, Waksman 1919, emend. Preobrazhenskaya 1966, ATCC 11635

Streptomyces (Actinomyces) erythreus var. speleomycini Sabo et Preobrazhenskaya 1962, B-23

Streptomyces (Actinomyces) erythrochromogenes Krainsky 1914, Waksman et Henrici 1948, NIHJ 209

Streptomyces (Actinomyces) eurocidicus Okami et al. 1954, NRRL 1676

Streptomyces (Actinomyces) eurythermus Corbaz et al. 1955, ISP 5014

Streptomyces (Actinomyces) felleus Lindenbein 1952, ISP 5130

Streptomyces (Actinomyces) filamentosus Okami et al. 1953, ISP 5022

Streptomyces (Actinomyces) fimbriatus Millard et Burr 1926, Waksman et Henrici 1948, ATCC 15051

Actinomyces (Streptomyces) flavidovirens var. fuscus Gause et al. 1957, INA 15719

Actinomyces (Streptomyces) flavidovirens Gause et al. 1957, INA 12287

Streptomyces (Actinomyces) flavochromogenes Krainsky 1914, Waksman et Henrici 1948, INA 743/54

Actinomyces (Streptomyces) flavofungini Sabo et Preobrazhenskaya 1962, holotype SA 9

Streptomyces (Actinomyces) flavopersicus Oliver et al. 1961, ISP 5093

Actinomyces (Streptomyces) flavotricini Gause et al. 1957, INA 11669/58

Streptomyces (Actinomyces) flavovirens Waksman 1919, Waksman et Henrici 1948, IMRU 3320

Streptomyces (Actinomyces) flavus Krainsky 1914, emend. Krassilnikov 1941, ATCC 3369

Streptomyces (Actinomyces) flocculus Duché 1934, NRRL B-1905

Actinomyces (Streptomyces) fluorescens Korenyako 1960, INMI 592

Streptomyces (Actinomyces) fradiae Waksman et Henrici 1948, ATCC 10745

Streptomyces (Actinomyces) fulvissimus Jensen 1930, Waksman et Henrici 1948, IPV 514

Actinomyces (Streptomyces) fulvoviridis Kuchaeva et al. 1960, INMI 607

Actinomyces (Streptomyces) fumanus Gause et al. 1957, INA 10256/54

Actinomyces (Streptomyces) fumosus Krassilnikov 1941, INA 4413-a

Streptomyces (Actinomyces) graminofaciens Charney et al. 1953, PSA 153

Streptomyces (Actinomyces) galilaeus Ettlinger et al. 1958, ATCC 14969

Streptomyces (Actinomyces) galbus Frommer 1959, PSA 208

Streptomyces (Actinomyces) gelaticus Waksman et Henrici 1948, Waksman ATCC 3323

Streptomyces (Actinomyces) gibsonii Erikson 1935, emend. Preobrazhenskaya 1966, INA 10226

Actinomyces (Streptomyces) globisporus var. tundromycini Kovalenkova 1957, INA 1078

Actinomyces (Streptomyces) globisporus var. flavofuscus Gause et al. 1957, INA 1140, 1803

Actinomyces (Streptomyces) globisporus var. caucasicus Gause et al. 1957, INA 13195/54

Actinomyces (Streptomyces) globisporus Krassilnikov 1941, INMI 2302

Streptomyces (Actinomyces) goshikiensis Niida (Kondo et al.) 1961, ISP 5190

Actinomyces (Streptomyces) graminearus Berestnev 1897, Gause et al. 1957, INA 13892

Streptomyces (Actinomyces) griseinus Waksman 1948, ISP 5047

Streptomyces (Actinomyces) griseocarneus Benedict et al. 1951, ISP 5004

Streptomyces (Actinomyces) griseofuscus Sakamoto et al. 1962, ISP 5191

Actinomyces (Streptomyces) griseoloalbus Gause et al. 1957, INA 1875/54

Streptomyces (Actinomyces) griseolus Waksman 1919, Waksman et Henrici 1948, INA 3950

Streptomyces (Actinomyces) griseoluteus Umezawa et al. 1951, ATCC 12768

Streptomyces (Actinomyces) griseoluteus Umezawa et al. 1951, NIHJ P-37

Streptomyces (Actinomyces) griseoruber Yamaguchi et Saburi 1955, ISP 5281

Actinomyces (Streptomyces) griseorubiginosus var. spiralis Gause et al. 1957, INA 7989

Actinomyces (Streptomyces) griseorubiginosus Gause et al. 1957, INA 7712

Actinomyces (Streptomyces) griseovariabilis Krassilnikov 1949, INA 1360
Streptomyces (Actinomyces) griseoviridis Anderson et al. 1956, ISP 5229, INA 6613
Streptomyces (Actinomyces) griseus Waksman et Henrici 1948, IMRU 3475
Streptomyces (Actinomyces) halstedii Waksman et Curtis 1916, Waksman et Henrici 1948, ATCC 10897
Streptomyces (Actinomyces) herbaricolor Kawato et Shinobu 1959, OUE 608
Actinomyces (Streptomyces) herbescens Krassilnikov et Egorova 1965, INMI 1252
Streptomyces (Actinomyces) hiroshimensis Shinobu 1955, INA 10204/54
Streptomyces (Actinomyces) hominis Bostroem 1890, emend. Waksman 1919, Waksman et Henrici 1948, ATCC 3008
Streptomyces (Actinomyces) humidus Nakazawa et Shibata 1956, ISP 5263
Streptomyces (Actinomyces) hygroscopicus Jensen 1931, Waksman et Henrici 1948, ATCC 10976
Streptomyces (Actinomyces) indigoferus Shinobu et Kawato 1960, ISP 5124
Actinomyces (Streptomyces) intermedius Wollenweber 1922, emend. Gause et al. 1957, INA 10976
Streptomyces (Actinomyces) kanamyceticus Okami et Umezawa 1957, ATCC 12853
Streptomyces (Actinomyces) kentuckensis Barr et German 1956, INA 8211
Actinomyces (Streptomyces) kurssanovii Gause et al. 1957, INA 7069a/54
Actinomyces (Streptomyces) lavendulae var. rubescens Sveshnikova 1966, INA 3801/55
Streptomyces (Actinomyces) lavendulae Waksman et Henrici 1948, INA 4518, 840
Actinomyces (Streptomyces) levis Sveshnikova et al. 1967, INA 9020
Streptomyces (Actinomyces) lilacinus Nakazawa et al. 1957, 4-216
Streptomyces (Actinomyces) lincolnensis var. lincolnensis Mason et al. 1962, holotype 2936
Streptomyces (Actinomyces) lipmanii Waksman et Curtis 1916, Waksman et Henrici 1948, ATCC 3331
Actinomyces (Streptomyces) litmocidini Gause et al. 1957, INA 1823/55
Actinomyces (Streptomyces) longisporoflavus Krassilnikov 1941, ISP 5165
Actinomyces (Streptomyces) longispororuber Krassilnikov 1941, INA 8173
Actinomyces (Streptomyces) longisporus Krassilnikov 1941, ISP 5166
Actinomyces (Streptomyces) luridus Krassilnikov et al. 1957, INMI III
Streptomyces (Actinomyces) luteolutescens Yen Hsun-chu 1957, 220
Streptomyces (Actinomyces) luteoverticillatus Shinobu 1956, holotype 486
Streptomyces (Actinomyces) lydicus DeBoer et al. 1955, ISP 5461
Actinomyces (Streptomyces) malachitorectus Preobrazhenskaya et al. 1964, INA 8554
Streptomyces (Actinomyces) macrosporeus Ettlinger et al. 1958, ETH 7534
Streptomyces (Actinomyces) machuensis Sawazaki et al. 1955, ISP 5221
Streptomyces (Actinomyces) melanochromogenes Tsai Yung-shen 1957, holotype 1779
Streptomyces (Actinomyces) microflavus Krainsky 1914, Waksman et Henrici 1948, ISP 5331
Streptomyces (Actinomyces) mitakaensis Arai et al. 1958, ATCC 15297
Streptomyces (Actinomyces) minoensis Nishimura 1961, ISP 5031
Streptomyces (Actinomyces) mirabilis Ruschmann 1952, ISP 5168
Streptomyces (Actinomyces) misakiensis Nakamura 1961, ISP 5222
Streptomyces (Actinomyces) michiganensis Corbaz et al. 1957, ETH 15229
Actinomyces (Streptomyces) mutabilis Gause et al. 1957, INA 472
Streptomyces (Actinomyces) murinus Frommer 1959, ISP 5091
Streptomyces (Actinomyces) naganishii Yamaguchi et Saburi 1955, NRRL 1816
Streptomyces (Actinomyces) narbonensis Corbaz et al. 1955, ETH 7346
Streptomyces (Actinomyces) netropsis Finlay et Sobin 1952, NRRL 2268
Actinomyces (Streptomyces) nigrescens Gause et al. 1957, INA 1800/54
Streptomyces (Actinomyces) nigrifaciens Waksman 1919, ISP 5071
Streptomyces (Actinomyces) niveoruber Ettlinger et al. 1958, ETH 17860
Streptomyces (Actinomyces) niveus Smith et al. 1956, ISP 5088
Streptomyces (Actinomyces) novaecaesareae Waksman et Curtis 1916, Waksman et Henrici 1948, CBS
Streptomyces (Actinomyces) odorifer Lachner-Sandoval 1898, Waksman 1961, ATCC 2646
Actinomyces (Streptomyces) oidiosporus Krassilnikov 1941, INA 4020

Actinomyces (Streptomyces) olivaceoviridis Gause et al. 1957, INA 11584a
Streptomyces (Actinomyces) olivaceus Waksman 1919, Waksman et Henrici 1948, ATCC 11626
Streptomyces (Actinomyces) olivoreticuli Arai et al. 1957, NRRL B-2091
Actinomyces (Streptomyces) olivoreticuli var. olivomycini Gause et al. 1962, INA 16749
Streptomyces (Actinomyces) olivochromogenes Waksman 1919, Waksman et Henrici 1948, ATCC 3336
Streptomyces (Actinomyces) ostreogriseus Ball et al. 1958, NCIB 8792
Streptomyces (Actinomyces) orientalis Pittenger et Brighan 1956, RSUA 130
Actinomyces (Streptomyces) pallidoviolaceus Sveshnikova et al. 1967, INA 10942
Streptomyces (Actinomyces) parvulus Waksman et Gregory 1954, IMRU 3677
Streptomyces (Actinomyces) parvus Krainsky 1914, Waksman et Henrici 1948, NRRL B-1455
Streptomyces (Actinomyces) phaeochromogenes Conn 1917, Waksman et Henrici 1948, NIHJ 108-A
Actinomyces (Streptomyces) prunicolor Gause et al. 1957, INA 9050
Streptomyces (Actinomyces) platensis Pittenger et Gottlieb 1954, ISP 5041
Streptomyces (Actinomyces) polychromogenes Hagemann et al. 1955, NCIB 8791
Actinomyces (Streptomyces) proteolyticus Sveshnikova 1966, INA 14013
Streptomyces (Actinomyces) praecox Millard et Burr 1926, Waksman et Henrici 1948, Bu-1.75
Streptomyces (Actinomyces) psammoticus Virgilio et Hengeller 1960, holotype
Streptomyces (Actinomyces) pseudogriseolus Okami et al. 1955, ATCC 12770
Streptomyces (Actinomyces) pseudovenezuelae Kutschaeva 1962, ISP 5212
Streptomyces (Actinomyces) purpeofuscus Yamaguchi 1955, ISP 5283
Streptomyces (Actinomyces) phaeoverticillatus Matsumae et al. 1964, B-2771
Streptomyces (Actinomyces) ramulosus Ettlinger et al. 1958, ISP 5100
Streptomyces (Actinomyces) resistomycificus Lindenbein 1952, ISP 5133
Streptomyces (Actinomyces) rimosus Sobin et al. 1950, NRRL 2234
Streptomyces (Actinomyces) rimosus f. paromomycini 1958, Brit. Pat. 14827
Streptomyces (Actinomyces) roseoflavus Arai 1951, INA 3489
Actinomyces (Streptomyces) roseofulvus Gause et al. 1957, INA 14535
Actinomyces (Streptomyces) roseofulvus var. tauricus Sveshnikova 1962, INA 323
Actinomyces (Streptomyces) roseolilacinus Gause et al. 1957, INA 14250
Actinomyces (Streptomyces) roseoviridis Gause et al. 1957, INA 3617
Streptomyces (Actinomyces) rutgersensis Waksman et Curtis 1916, Waksman et Henrici 1948, ISP 5077
Actinomyces (Streptomyces) rutilans Sveshnikova et al. 1967, INA 2245
Streptomyces (Actinomyces) salmonicida Rucker 1949, NRRL B-1472
Streptomyces (Actinomyces) scabies Waksman et Henrici 1948, IFO 3111
Streptomyces (Actinomyces) sioyaensis Nishimura et al. 1961, ISP 5032
Streptomyces (Actinomyces) spiroverticillatus Shinobu 1958, OUE 508
Actinomyces (Streptomyces) streptomycini Krassilnikov 1955, INMI 1780
Streptomyces (Actinomyces) sulphureus Rivolta 1882, emend. Gasperini 1894, Waksman, NRRL B-1627
Actinomyces (Streptomyces) syringini Gause et al. 1957, INA 2020
Actinomyces (Streptomyces) ukrainicus Maximova 1966, INA 2626
Actinomyces (Streptomyces) umbrinus Gause et al. 1957, INA 1703/53
Actinomyces (Streptomyces) tauricus Sveshnikova 1966, INA 8173
Streptomyces (Actinomyces) tendae Ettlinger et al. 1958, ETH 14077
Streptomyces (Actinomyces) thioluteus Okami 1952, holotype 26-A
Actinomyces (Streptomyces) toxytricini Gause et al. 1957, INA 13887/54
Streptomyces (Actinomyces) tubercidicus Nakamura 1961, ISP 5261
Streptomyces (Actinomyces) varsoviensis Kuryłowicz 1954, ATCC 14631
Streptomyces (Actinomyces) venezuelae Ehrlich et al. 1948, ATCC 10595
Actinomyces (Streptomyces) venezuelae var. spiralis Gause et al. 1957, INA 11686
Streptomyces (Actinomyces) vinaceus Mayer et al. 1951, NRRL B-2585
Streptomyces (Actinomyces) violaceoniger Waksman et Curtis 1916, INA 9930

Actinomyces (Streptomyces) violaceorectus Gause et al. 1957, INA 14046

Actinomyces (Streptomyces) violaceus Gasperini 1894, emend. Krassilnikov 1941, emend. Maximova et Preobrazhenskaya 1966, INA 5551/54

Actinomyces (Streptomyces) violaceus var. rubescens Gause et al. 1957, INA 8155

Streptomyces (Actinomyces) violochromogenes Artamonova 1960, ISP 5207

Streptomyces (Actinomyces) viridodiastaticus Baldacci et al. 1955, ISP 5249

Streptomyces (Actinomyces) virocidus (virusinus) Kuchaeva et al. 1962, INMI 1609

Streptomyces (Actinomyces) viridifaciens US patent 1955, ISP 5239

Actinomyces (Streptomyces) vulgaris Nikitina et al. 1960, INMI 1034

Streptomyces (Actinomyces) willmorei Erikson 1935, Waksman et Henrici 1948, ISP 5459

Streptomyces (Actinomyces) xanthochromogenes Arichima 1956, ISP 5111

Streptomyces (Actinomyces) xanthophaeus Lindenbein 1952, ISP 5134

Streptomyces (Actinomyces) zaomyceticus Hinuma 1954, ISP 5196

Index of strains described in the atlas

References

BALDACCI E., GREIN A.: Esame della forma delle spore di attinomiceti al microscopio elettronico e loro valuazione ai fini di una classificazione. *Giorn. Microbiol.*, 1, 28—34, 1955

DIETZ A., MATHEWS J.: Taxonomy by carbon replication. I. An examination of Streptomyces hygroscopicus. *Appl. Microbiol.*, 10, 258—263, 1962

ECHLIN P.: Intra-cytoplasmic membranous inclusions in the blue-green alga, Anacystis nidulans. *Archiv. Mikrobiol.*, 49, 267—274, 1964

ETTLINGER L., CORBAZ R., HÜTTER R.: Zur Artenteilung der Gattung Streptomyces Waksman et Henrici. *Arch. Mikrobiol.*, 31, 326—358, 1958

FLAIG W., KÜSTER E., BEUTELSPACHER H., SCHLICHTING-BAUER I., POLITT-RUNGE W., KURZ R.: Elektronenmikroskopische Untersuchungen an Sporen verschiedener Streptomyceten. *Zentr. Bakteriol. Parasitenk.* Abt. II, 108, 376—382, 1955

GAUSE G. F., PREOBRAZHENSKAYA T. P., KUDRINA E. S., BLINOV N. O., RYABOVA I. D., SVESHNIKOVA M. A.: *Problems of classifications of Actinomycetes-antagonists* (in Russian). Medgiz Moscow, 1957

GREIN A.: Una tecnica per l'osservazione di attinomiceti al microscopio elettronico. *Riv. il lab. scientifico*, No. 3, 1955

GYLLENBERG H. G., WOŹNICKA W., KURYŁOWICZ W.: Application of factor analysis in microbiology. 3. A study of the "Yellow series" of Streptomycetes. *Ann. Academiae Scientiarum Fennicae*, Series A, IV Biologica, 114, 1—14, 1967

HOPWOOD D. A., GLAUERT A. M.: Electron microscope observation on the surface structures of Streptomyces violaceoruber. *J. Gen. Microbiol.*, 26, 325—329, 1961

KRISS A. E., RUKINA E. A., ISSAIEV B. M.: Electron microscopic studies on the structure of Actinomycetes (in Russian). *Mikrobiologiya*, 14, 172—176, 1945

KURYŁOWICZ W., WOŹNICKA W., PASZKIEWICZ A., MALINOWSKI K.: Application of numeric taxonomy in Streptomycetes. H. Prauser (ed.). *The Actinomycetales.* G. Fischer Verlag, Jena, 1970.

KÜSTER E.: Beitrag zur Genese und Morphologie der Streptomycetensporen. *Atti VI Intern. Congr. Microbiol.* 1, 114—116, 1953

LECHEVALIER H. A., HOLBERT P. E.: Electron microscopic observation on the sporangial structure of a strain of Actinoplanes. *J. Bacteriol.*, 89, 217—222, 1965

LECHEVALIER H. A., LECHEVALIER M. P., HOLBERT P. E.: Electron microscopic observation on the sporangial structure of strains of Actinoplanaceae. *J. Bacteriol.*, 92, 1228—1235, 1966

OKAMI Y.: *Atlas of electron-micrograms of the Actinomycetes.* The Society for Actinomycetes, Japan, 1965

PREOBRAZHENSKAYA T. P., KUDRINA J. S., SVESHNIKOVA M. A., MAXIMOVA T. S.: The use of electron microscopy of spores in the systhematic of Actinomyces (in Russian). *Mikrobiologiya*, 28, 623—627, 1959

PREOBRAZHENSKAYA T. P., KUDRINA J. S., MAXIMOWA T. S., SVESHNIKOVA M. A., BOYARSKAYA R. V.: Studies in electron microscopy of spores of various Actinomycetes species (in Russian). *Mikrobiologiya*, 29, 51—55, 1960

PREOBRAZHENSKAYA T. P., MAXIMOVA T. S., LUKYANOVICH W. M., EVKO E. I.: The use of the carbon replica method for an electron-microscopic examination of the surface of Actinomyces spores (in Russian). *Mikrobiologiya*, 34, 519—523, 1965

RANCOURT M., LECHEVALIER H. A.: Electron microscopic observation of the sporangial structure of an actinomycete, Microellobosporia flavea. *J. Gen. Microbiol.*, **31**, 495—498, 1963

RANCOURT M., LECHEVALIER H. A.: Electron microscopic study of the formation of spiny conidia in species of Streptomyces. *Canad. J. Microbiol.*, **10**, 311—316, 1964

SABATINI D. D., BENSCH K., BARRNETT R. J.: Cytochemistry and electron microscopy. The preservation of cellular ultrastructure and enzymatic activity by aldehyde fixation. *J. Cell Biology*, **17**, 19—58, 1963

SHIRLING E. B., GOTTLIEB D.: *Methods Manual,* International Cooperative Project for Description and Deposition of Type Cultures of Streptomycetes, National Science Foundation, USA, 1964

TRESNER H. D., DAVIES M. C., BACKUS E. J.: Electron microscope studies of spore morphology in the genus Streptomyces. *Bacteriol. Proc. Abstr.* 60 th Ann. Mtg., p. 53, 1960

TRESNER H. D., DAVIES M. C., BACKUS E. J.: Electron microscope of Streptomyces spore morphology and its role in species differentiation. *J. Bacteriol.*, **81**, 70—80, 1961

WILLIAMS S. T., DAVIES F. L: Use of a scanning electron microscope for the examination of Actinomycetes. *J. Gen. Microbiol.*, **48**, 171—177, 1967

WOŹNICKA W.: Trials of classification of the ''Yellow series'' of Actinomycetes. I. Taxonomic studies. *Exp. Med. and Microbiol.*, **19**, 10—22, 1967

WOŹNICKA W.: Trials of classification of the ,,Yellow series'' of Actinomycetes. II. Taxonometric studies. *Exp. Med. and Microbiol.*, **19**, 23—29, 1967

YAJIMA Y., AMANO S., NIIDA T.: Electron microscopic study of spores of Streptomyces. *Scientific Reports of Meiji Seika Kaisha*, **7**, 41—47, 1965

This book, designed by K. Blichewicz,
is printed and bound in Poland at
Drukarnia Techniczno-Naukowa im. Rewolucji Październikowej
Warsaw

The half-tone blocks made by
Dom Słowa Polskiego
Warsaw